Climatology: Principles, Models and Applications

Climatology: Principles, Models and Applications

Salvador Poole

SYRAWOOD
PUBLISHING HOUSE
New York

Published by Syrawood Publishing House,
750 Third Avenue, 9th Floor,
New York, NY 10017, USA
www.syrawoodpublishinghouse.com

Climatology: Principles, Models and Applications
Salvador Poole

International Standard Book Number: 978-1-64740-023-1 (Hardback)

Cataloging-in-Publication Data

Climatology : principles, models and applications / Salvador Poole.
   p. cm.
Includes bibliographical references and index.
ISBN 978-1-64740-023-1
1. Climatology. 2. Climatology--Models. 3. Atmospheric science. I. Poole, Salvador.
QC981 .C55 2020
551.6--dc23

TABLE OF CONTENTS

PREFACE

The branch of science which studies weather conditions over a period of time is known as climatology. It is a branch of atmospheric sciences. Climatology is also involved in the study of dynamics of the weather and climate system. The time period over which these weather events are studied can stretch from a few years to millennium. There are a number of sub-disciplines within this field such as boundary-layer climatology, paleoclimatology and paleotempestology. The study of climate indices is an important part of climatology. These indices are used to represent the important elements of climate. Some of the important phenomena studied within this field are circulation patterns, heat transfer and atmospheric boundary layer. Climatology is an upcoming field of science that has undergone rapid development over the past few decades. The topics covered herein offer the readers new insights in this field. The book is appropriate for students seeking detailed information in this area as well as for experts.

To facilitate a deeper understanding of the contents of this book a short introduction of every chapter is written below:

Chapter 1- A long-term averaged weather is defined as climate. The scientific study of the weather conditions over a period of time is referred to as climatology. It depends upon various factors such as Earth-sun relationship, climate variability, effects of topography, climate modeling, etc. This chapter has been carefully written to provide an easy understanding of the varied facts of climatology.

Chapter 2- There are different types of climate depending upon the terrain such as tropical climate, dry climate, temperate climate, continental climate, polar climate, etc. The aim of this chapter is to explore the different types of climate. These topics are crucial for a complete understanding of climatology.

Chapter 3- The primary factors affecting the climate include humidity, atmospheric circulation, ocean and wind currents, atmospheric pressure, winds, precipitation, temperature, geography, latitude, neutral phase, etc. This chapter closely examines these phenomena of climatology to provide an extensive understanding of the subject.

Chapter 4- Long-term rise in the average temperature of the Earth's climate system is referred to as global warming. It consists of various studies such as impacts of climate change on ecosystems and human health, air pollution and climate change, etc. Climate change and global warming is an interdisciplinary subject which makes it essential to understand its related studies.

Chapter 5- In order to completely understand climatology it is necessary to understand the impacts of human activities on climate. The following chapter elucidates the varied impacts associated with this area of study. Human activities such as deforestation, biomass burning, carbon dioxide emission, etc., have a negative impact on the climate.

Chapter 6- Diverse instruments are used for the study of climate. This chapter delves into the usage of equipment such as thermometer, barometer, hygrometer, anemometer, pyranometer, rain gauge, disdrometer, transmissometer, ceilometer, etc. to provide an extensive understanding of this subject.

I owe the completion of this book to the never-ending support of my family, who supported me throughout the project.

Salvador Poole

Chapter 1

Understanding Climatology

A long-term averaged weather is defined as climate. The scientific study of the weather conditions over a period of time is referred to as climatology. It depends upon various factors such as Earth-sun relationship, climate variability, effects of topography, climate modeling, etc. This chapter has been carefully written to provide an easy understanding of the varied facts of climatology.

Climate Classification

Climate classification is the formalization of systems that recognize, clarify, and simplify climatic similarities and differences between geographic areas in order to enhance the scientific understanding of climates. Such classification schemes rely on efforts that sort and group vast amounts of environmental data to uncover patterns between interacting climatic processes. All such classifications are limited since no two areas are subject to the same physical or biological forces in exactly the same way. The creation of an individual climate scheme follows either a genetic or an empirical approach.

The climate of an area is the synthesis of the environmental conditions (soils, vegetation, weather, etc.) that have prevailed there over a long period of time. This synthesis involves both averages of the climatic elements and measurements of variability (such as extreme values and probabilities). Climate is a complex, abstract concept involving data on all aspects of Earth's environment. As such, no two localities on Earth may be said to have exactly the same climate.

Nevertheless, it is readily apparent that, over restricted areas of the planet, climates vary within a limited range and that climatic regions are discernible within which some uniformity is apparent in the patterns of climatic elements. Moreover, widely separated areas of the world possess similar climates when the set of geographic relationships occurring in one area parallels that of another. This symmetry and organization of the climatic environment suggests an underlying worldwide regularity and order in the phenomena causing climate (such as patterns of incoming solar radiation, vegetation, soils, winds, temperature, and air masses). Despite the existence of such underlying patterns, the creation of an accurate and useful climate scheme is a daunting task.

First, climate is a multidimensional concept, and it is not an obvious decision as to which of the many observed environmental variables should be selected as the basis of the classification. This choice must be made on a number of grounds, both practical and theoretical. For example, using too many different elements opens up the possibilities that the classification will have too many categories to be readily interpreted and that many of the categories will not correspond to real climates. Moreover, measurements of many of the elements of climate are not available for large areas of the world or have been collected for only a short time. The major exceptions are soil, vegetation, temperature, and precipitation data, which are more extensively available and have been recorded for extended periods of time.

The choice of variables also is determined by the purpose of the classification (such as to account for distribution of natural vegetation, to explain soil formation processes, or to classify climates in terms of human comfort). The variables relevant in the classification will be determined by this purpose, as will the threshold values of the variables chosen to differentiate climatic zones.

A second difficulty results from the generally gradual nature of changes in the climatic elements over Earth's surface. Except in unusual situations due to mountain ranges or coastlines, temperature, precipitation, and other climatic variables tend to change only slowly over distance. As a result, climate types tend to change imperceptibly as one moves from one locale on Earth's surface to another. Choosing a set of criteria to distinguish one climatic type from another is thus equivalent to drawing a line on a map to distinguish the climatic region possessing one type from that having the other. While this is in no way different from many other classification decisions that one makes routinely in daily life, it must always be remembered that boundaries between adjacent climatic regions are placed somewhat arbitrarily through regions of continuous, gradual change and that the areas defined within these boundaries are far from homogeneous in terms of their climatic characteristics.

Most classification schemes are intended for global- or continental-scale application and define regions that are major subdivisions of continents hundreds to thousands of kilometres across. These may be termed macroclimates. Not only will there be slow changes (from wet to dry, hot to cold, etc.) across such a region as a result of the geographic gradients of climatic elements over the continent of which the region is a part, but there will exist mesoclimates within these regions associated with climatic processes occurring at a scale of tens to hundreds of kilometres that are created by elevation differences, slope aspect, bodies of water, differences in vegetation cover, urban areas, and the like. Mesoclimates, in turn, may be resolved into numerous microclimates, which occur at scales of less than 0.1 km (0.06 mile), as in the climatic differences between forests, crops, and bare soil, at various depths in a plant canopy, at different depths in the soil, on different sides of a building, and so on.

These limitations notwithstanding, climate classification plays a key role as a means of generalizing the geographic distribution and interactions among climatic elements, of identifying mixes of climatic influences important to various climatically dependent phenomena, of stimulating the search to identify the controlling processes of climate, and, as an educational tool, to show some of the ways in which distant areas of the world are both different from and similar to one's own home region.

Approaches to Climatic Classification

The earliest known climatic classifications were those of Classical Greek times. Such schemes generally divided Earth into latitudinal zones based on the significant parallels of 0°, 23.5°, and 66.5° of latitude (that is, the Equator, the Tropics of Cancer and Capricorn, and the Arctic and Antarctic circles, respectively) and on the length of day. Modern climate classification has its origins in the mid-19th century, with the first published maps of temperature and precipitation over Earth's surface, which permitted the development of methods of climate grouping that used both variables simultaneously.

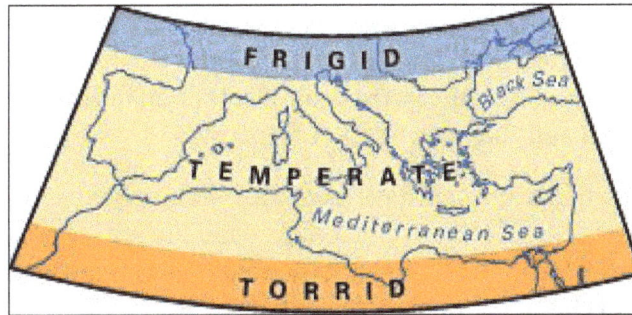

Map of climatic zones.

Many different schemes of classifying climate have been devised (more than 100), but all of them may be broadly differentiated as either empiric or genetic methods. This distinction is based on the nature of the data used for classification. Empirical methods make use of observed environmental data, such as temperature, humidity, and precipitation, or simple quantities derived from them (such as evaporation). In contrast, a genetic method classifies climate on the basis of its causal elements, the activity and characteristics of all factors (air masses, circulation systems, fronts, jet streams, solar radiation, topographic effects, and so forth) that give rise to the spatial and temporal patterns of climatic data. Hence, while empirical classifications are largely descriptive of climate, genetic methods are (or should be) explanatory. Unfortunately, genetic schemes, while scientifically more desirable, are inherently more difficult to implement because they do not use simple observations. As a result, such schemes are both less common and less successful overall. Moreover, the regions defined by the two types of classification schemes do not necessarily correspond; in particular, it is not uncommon for similar climatic forms resulting from different climatic processes to be grouped together by many common empirical schemes.

Genetic Classifications

Genetic classifications group climates by their causes. Among such methods, three types may be distinguished: (1) those based on the geographic determinants of climate, (2) those based on the surface energy budget, and (3) those derived from air mass analysis.

In the first class are a number of schemes (largely the work of German climatologists) that categorize climates according to such factors as latitudinal control of temperature, continentality versus ocean-influenced factors, location with respect to pressure and wind belts, and effects of mountains. These classifications all share a common shortcoming: they are qualitative, so that climatic regions are designated in a subjective manner rather than as a result of the application of some rigorous differentiating formula.

An interesting example of a method based on the energy balance of Earth's surface is the 1970 classification of Werner H. Terjung, an American geographer. His method utilizes data for more than 1,000 locations worldwide on the net solar radiation received at the surface, the available energy for evaporating water, and the available energy for heating the air and subsurface. The annual patterns are classified according to the maximum energy input, the annual range in input, the shape of the annual curve, and the number of months with negative magnitudes (energy deficits). The combination of characteristics for a location is represented by a label consisting of several letters with defined meanings, and regions having similar net radiation climates are mapped.

Probably the most extensively used genetic systems, however, are those that employ air mass concepts. Air masses are large bodies of air that, in principle, possess relatively homogeneous properties of temperature, humidity, etc., in the horizontal. Weather on individual days may be interpreted in terms of these features and their contrasts at fronts.

Two American geographer-climatologists have been most influential in classifications based on air mass. In 1951 Arthur N. Strahler described a qualitative classification based on the combination of air masses present at a given location throughout the year. Some years later (1968 and 1970) John E. Oliver placed this type of classification on a firmer footing by providing a quantitative framework that designated particular air masses and air mass combinations as "dominant," "subdominant," or "seasonal" at particular locations. He also provided a means of identifying air masses from diagrams of mean monthly temperature and precipitation plotted on a "thermohyet diagram," a procedure that obviates the need for less common upper-air data to make the classification.

Empirical Classifications

Most empirical classifications are those that seek to group climates based on one or more aspects of the climate system. While many such phenomena have been used in this way, natural vegetation stands out as one of prime importance. The view held by many climatologists is that natural vegetation functions as a long-term integrator of the climate in a region; the vegetation, in effect, is an instrument for measuring climate in the same way that a thermometer measures temperature.

Wladimir Köppen, a German botanist-climatologist, developed the most popular (but not the first) of these vegetation-based classifications. His aim was to devise formulas that would define climatic boundaries in such a way as to correspond to those of the vegetation zones that were being mapped for the first time during his lifetime. Köppen published his first scheme in 1900 and a revised version in 1918. He continued to revise his system of classification until his death in 1940. Other climatologists modified portions of Köppen's procedure on the basis of their experience in various parts of the world.

Köppen's classification is based on a subdivision of terrestrial climates into five major types, which are represented by the capital letters A, B, C, D, and E. Each of these climate types except for B is defined by temperature criteria. Type B designates climates in which the controlling factor on vegetation is dryness (rather than coldness). Aridity is not a matter of precipitation alone but is defined by the relationship between the precipitation input to the soil in which the plants grow and the evaporative losses. Since evaporation is difficult to evaluate and is not a conventional measurement at meteorological stations, Köppen was forced to substitute a formula that identifies aridity in terms of a temperature-precipitation index (that is, evaporation is assumed to be controlled by temperature). Dry climates are divided into arid (BW) and semiarid (BS) subtypes, and each may be differentiated further by adding a third code, for warm (h) or cold (k).

Temperature defines the other four major climate types. These are subdivided, with additional letters again used to designate the various subtypes. Type A climates, the warmest, are differentiated on the basis of the seasonality of precipitation: Af (no dry season), Am (short dry season), or Aw (winter dry season). Type E climates, the coldest, are conventionally separated into tundra (ET) and snow/ice climates (EF). The midlatitude C and D climates are given a second letter, f (no dry season) or w (winter dry) or s (summer dry), and a third symbol—a, b, c, or d (the last subclass exists only for D climates)—indicating the warmth of the summer or the coldness of the winter. Although Köppen's classification did not consider the uniqueness of highland climate regions, the highland climate category, or H climate, is sometimes added to climate classification systems to account for elevations above 1,500 metres (about 4,900 feet). The table gives the specific criteria for the Köppen-Geiger-Pohl system of 1953.

Classification of major climatic types according to the modified Köppen-Geiger scheme			
letter symbol			
1st	2nd	3rd	criterion
A			temperature of coolest month 18 °C or higher
	f		precipitation in driest month at least 60 mm
	m		precipitation in driest month less than 60 mm but equal to or greater than 100 − $(r/25)$[1]
	w		precipitation in driest month less than 60 mm and less than 100 − $(r/25)$
B[2]			70% or more of annual precipitation falls in the summer half of the year and r less than 20t + 280, or 70% or more of annual precipitation falls in the winter half of the year and r less than 20t, or neither half of the year has 70% or more of annual precipitation and r less than 20t + 140[3]
	W		r is less than one-half of the upper limit for classification as a B type
	S		r is less than the upper limit for classification as a B type but is more than one-half of that amount
		h	t equal to or greater than 18 °C
		k	t less than 18 °C
C			temperature of warmest month greater than or equal to 10 °C, and temperature of coldest month less than 18 °C but greater than −3 °C
	s		precipitation in driest month of summer half of the year is less than 30 mm and less than one-third of the wettest month of the winter half
	w		precipitation in driest month of the winter half of the year less than one-tenth of the amount in the wettest month of the summer half
	f		precipitation more evenly distributed throughout year; criteria for neither s nor w satisfied
		a	temperature of warmest month 22 °C or above
		b	temperature of each of four warmest months 10 °C or above but warmest month less than 22 °C
		c	temperature of one to three months 10 °C or above but warmest month less than 22 °C
D			temperature of warmest month greater than or equal to 10 °C, and temperature of coldest month −3 °C or lower
	s		same as for type C
	w		same as for type C
	f		same as for type C
		a	same as for type C
		b	same as for type C

		c	same as for type C
		d	temperature of coldest month less than −38 °C (d designation then used instead of a, b, or c)
E			temperature of warmest month less than 10 °C
	T		temperature of warmest month greater than 0 °C but less than 10 °C
	F		temperature of warmest month 0 °C or below
H[4]			temperature and precipitation characteristics highly dependent on traits of adjacent zones and overall elevation—highland climates may occur at any latitude

In the formulas above, r is average annual precipitation total (mm), and t is average annual temperature (°C). All other temperatures are monthly means (°C), and all other precipitation amounts are mean monthly totals (mm).Any climate that satisfies the criteria for designation as a B type is classified as such, irrespective of its other characteristics. The summer half of the year is defined as the months April–September for the Northern Hemisphere and October–March for the Southern Hemisphere. Most modern climate schemes consider the role of altitude.

The Köppen classification has been criticized on many grounds. It has been argued that extreme events, such as a periodic drought or an unusual cold spell, are just as significant in controlling vegetation distributions as the mean conditions upon which Köppen's scheme is based. It also has been pointed out that factors other than those used in the classification, such as sunshine and wind, are important to vegetation. Moreover, it has been contended that natural vegetation can respond only slowly to environmental change, so that the vegetation zones observable today are in part adjusted to past climates. Many critics have drawn attention to the rather poor correspondence between the Köppen zones and the observed vegetation distribution in many areas of the world. In spite of these and other limitations, the Köppen system remains the most popular climatic classification in use today.

A major contribution to climate grouping was made by the American geographer-climatologist C. Warren Thornthwaite in 1931 and 1948. He first used a vegetation-based approach that made use of the derived concepts of temperature efficiency and precipitation effectiveness as a means of specifying atmospheric effects on vegetation. His second classification retained these concepts in the form of a moisture index and a thermal efficiency index but radically changed the classification criteria and rejected the idea of using vegetation as the climatic integrator, attempting instead to classify "rationally" on the basis of the numerical values of these indices. His 1948 scheme is encountered in many climatology texts, but it has not gained as large a following among a wide audience as the Köppen classification system has, perhaps because of its complexity and the large number of climatic regions it defines.

While vegetation-based climate classifications could be regarded as having relevance to human activity through what they may indicate about agricultural potential and natural environment, they cannot give any sense of how human beings would feel within the various climate types. Terjung's 1966 scheme was an attempt to group climates on the basis of their effects on human comfort. The classification makes use of four physiologically relevant parameters: temperature, relative humidity, wind speed, and solar radiation. The first two are combined in a comfort index to express atmospheric conditions in terms perceived as extremely hot, hot, oppressive, warm, comfortable, cool, keen, cold, very cold, extremely cold, and ultra cold. Temperature, wind speed, and solar radiation are combined in a wind effect index expressing the net effect of wind chill (the cooling power of wind on exposed surfaces) and addition of heat to the human body by solar radiation.

These indices are combined for different seasons in different ways to express how humans feel in various geographic areas on a yearly basis. Terjung visualized that his classification would find applicability in medical geography, climatological education, tourism, housing, and clothing and as a general analytical tool.

Many other specialized empirical classifications have been devised. For example, there are those that differentiate between types of desert and coastal climates, those that account for different rates of rock weathering or soil formation, and those based on the identification of similar agricultural climates.

The Climate System

The climate system is extensive, multi-faceted, and all-pervasive. It has been defined as encompassing components of the atmosphere, hydrosphere, biosphere, geosphere, and cryosphere. Its space scales range from points to the planet, and the timeframes of interest to climate scientists extend from seconds to eons. Over the longest timescales, the Earth's climate has been remarkably stable: liquid water has been present on the planet's surface since at least 3.8 billion years ago. Glacial epochs have occurred throughout climate history prompting the build-up of masses of land-ice and hence the drawing down of sea levels globally. Disturbances from the mean, from catastrophes to hiccups, can be recognized in the climate record. These include periods when particulates, some caused by meteorite impacts and others from volcanos, dramatically cooled temperatures, and when a biophysically derived build-up of greenhouse gases warmed the Earth.

Future climates will be susceptible to the same internal and external forces as past climates. In addition, human activities are known to be modifying the atmospheric composition and the continental surface in ways that are likely to modify the pace and, perhaps, the direction of natural climatic change. Human impacts on climate differ from many natural processes because of the speed of the changes.

Climate models have been developed to try to characterize all the important features of the Earth's climate system. Different model types are intended to be better suited to specific aspects of climate change. Greenhouse climate prediction has been conducted with global climate models; integrated assessment models are being developed to assess the interactions of biospheric changes and climate shifts; and simpler energy budget models have been employed to explain the Milankovitch cycling of climate in response to orbital changes of the Earth around the Sun.

A hospitable climate is fundamental to sustainable development on Earth. Understanding climate sensitivities to internal and external changes and its reciprocal, biospheric, societal, and technological sensitivity to climate are crucially important to sustaining life systems on Earth.

Earth-sun Relationship

Figure below shows that the orbit of the Earth about the sun is not circular. The path is elongated or ellipitcal. This means that the distance from the Earth to the sun varies through the year. Two special events are depicted in the diagram. Aphelion (July 4) is when the Earth is as far away from the sun as it ever gets. Perihelion (Jan. 3) is when the Earth is as close to the sun as it ever gets. Note that these events do not correspond to the coldest and hottest months for us in the Northern Hemisphere. The purpose of this is to show that distance from the sun has nothing to do with seasons.

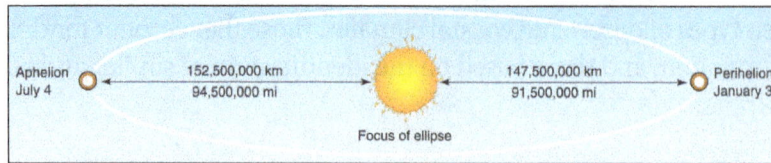

Figure looks rather complicated. It does, however, reveal some very important facts about the Earth and its orbit abound the sun. First note the purpleish rectangle. This represents the plane of the Earth's orbit about the sun or the Plane of the Ecliptic. We now want to measure the orientation of the Earth with respect to the plane of its orbit, the plane of the ecliptic. Now note the orange rectangle which represents the plane of the equator. We can clearly see that the two planes do not coincide. That is to say, the Earth is tilted with respect to the plane of the ecliptic. Figure also shows the Earth's axis of rotation. If the Earth were not tilted with respect to the plane of the ecliptic, then there would be a right angle (90°) between the axis and the plane of the ecliptic. Note that the axis is shy of 90° by 23°30'. This deviation, or tilt, is called Inclination. We will find that this inclination is vital for seasons on Earth.

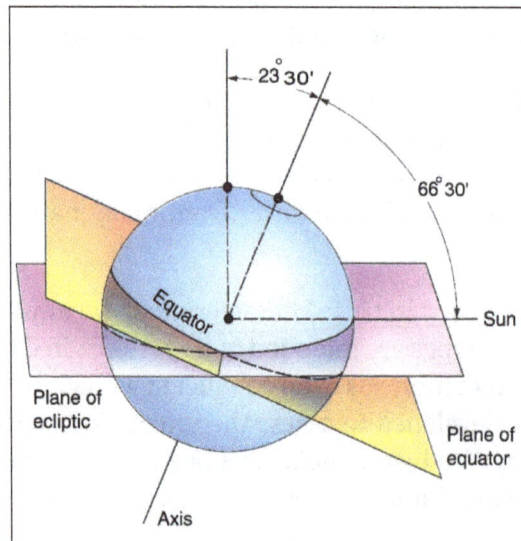

Additional:

- The spinning of the Earth about its axis is called Rotation.

- One rotation takes about 24 hours or 1 day.

Figure reveals two more important parts of the seasons story. First note that 50% of the Earth is in daylight and 50% is in darkness. This is always the case for the whole Earth, but equal parts of each hemisphere may not be in daylight and darkness. The dividing line between day and night is called the Circle of Illumination. The orientation of the circle of illumination changes with the seasons. Note in Figure that the circle of illumination does not pass through the poles. Look carefully and you will see that more of the Northern Hemisphere is in daylight than in darkness which means that the day is much longer than the night! What is important here is that the changing orientation of the circle of illumination alters the lengths of daylight and nighttime hours.

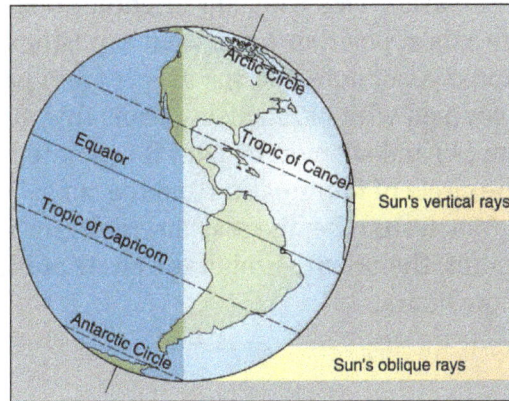

The second major concept shown in Figure is the Subsolar Point. The subsolar point is the latitude on the Earth's surface where the sun's rays strike at a 90° angle which is the highest possible solar angle. Figure shows a special event when the subsolar point is as far north as it ever gets, the Tropic of Cancer. The subsolar point is where the sun's rays are most direct and, therefore, most concentrated. The concentration of the solar energy heats the surface. Important rules emerge from this fact:

- When the subsolar point is as far north as it can go, it is the Northern Hemisphere's Summer.

- When the subsolar point is as far south as it can go, it is the Northern Hemisphere's Winter.

Figure is a view of the Earth from space showing the circle of illumination. Again, you can see that half of the planet is all ways in darkness and half is in daylight. The amounts of the northern and southern hemispheres in daylight and darkness, however, may NOT be equal. Read on and try to answer a question about this diagram posed below.

Figure shows the position of the Earth relative to the sun at four times of the year. You can see that the orbit is elliptical, as described earlier, and that the Earth exhibits a tilt (inclination)

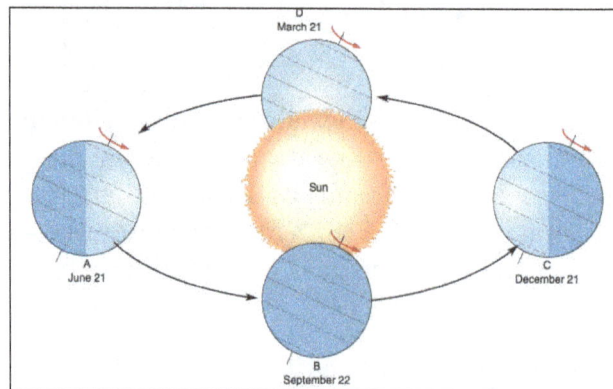

relative to the plane of its orbit around the sun (plane of the ecliptic). Figure also shows how the circle of illumination changes through the year. There is one final element that this figure shows that has a direct affect on seasons. Note the orientation of the Earth's axis. Do you see that the North Pole is always pointing in the same direction in space? The North Pole is always pointing at the "North Star" (Polaris). This constant orientation of the Earth's axis in space is called Parallelism. Look at the axis at position A and then at position C. Do you see that the axis is parallel in these two positions? Also, note that the axis is again parallel at positions B and D. The inclination of the Earth coupled with parallelism means that at one time of year the North Pole is pointed toward the sun (A) and six months later it is pointed away (C). This shift from A to C and back again causes the circle of illumination and the subsolar point to move and for the planet to experience seasons. When studying the seasons, make sure to note the tilt of the Earth, the position of the subsolar point, the orientiation of the circle of illumination, and the relative lenths of daylight and nighttime hours.

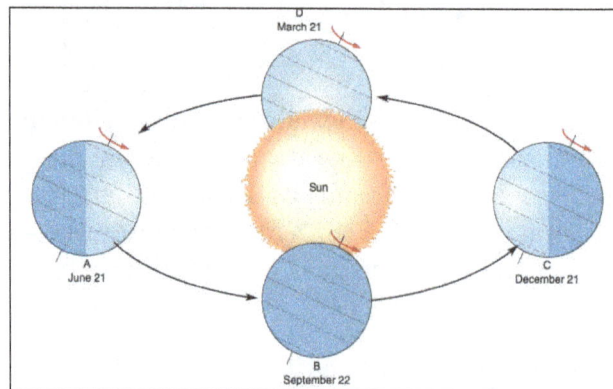

II Seasons:

Let's begin talking about seasons at March 21. At this point in time, the axis is neither pointed toward nor away from the sun. This causes the subsolar point to fall on the equator. The circle of illumination also passes through both poles making daylight and nighttime hours equal. When daylight and nighttime hours are equal, the event is called an Equinox. We, in the Northern Hemisphere, call March 21 the Vernal Equinox.

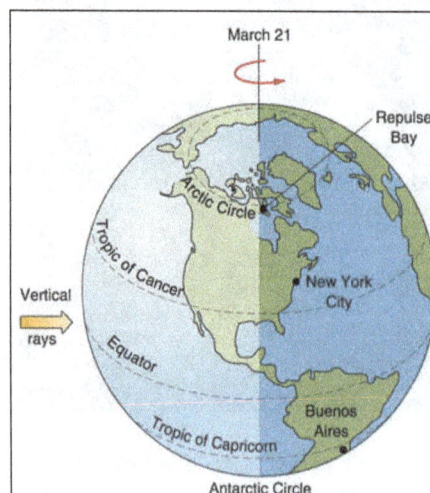

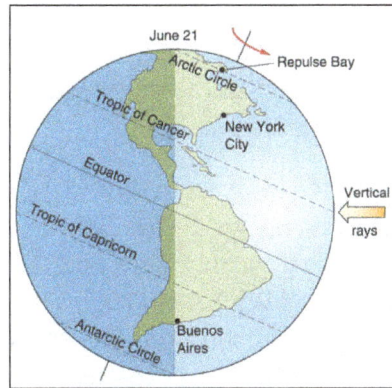

Three months later we arrive at June 21 (position A in Figure, and in Figure). Here the inclination of the Earth points the Northern Hemisphere toward the sun. This causes the subsolar point to be as far north as it ever goes (23°30' N), the Tropic of Cancer. The circle of illumination doesn't pass through both poles making daylight and nighttime hours differ to the extreme. Note that more of the Northern hemisphere is in daylight than in darkness. This represents the Northern Hemisphere's longest day of the year or the Summer Solstice. June 21 is also the shortest day in the Southern Hemisphere or their Winter Solstice. Since seasons are hemisphere specific, the June 21 event is called the June Solstice. Note that strange things happen on the June Solstice. Figure shows that Repulse Bay will not get rotated into darkness on this day. Anywhere on Repulse Bay's latitude will experience 24 hours of daylight. This latitude is 23°30' from the North Pole or at a latitude of 66°30' N. This is called the Arctic Circle. The Antarctic Circle, at 66°30' S experiences 24 hours of darkness on the June solstice.

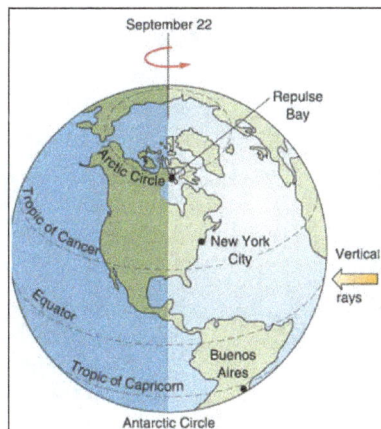

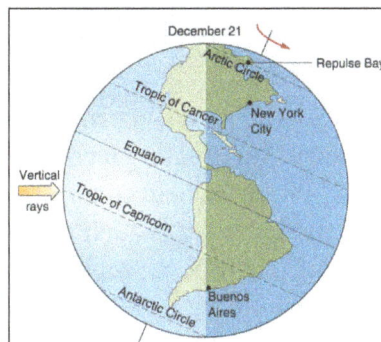

By September 22 (position B in Figure) the Earth is no longer pointed toward or away from the sun, and the subsolar point has returned to the Equator. The circle of illumination again passes through both poles making daylight and nighttime hours equal. This is the second equinox know as the Autumnal Equinox in the Northern Hemisphere.

On December 21, the north pole is pointed away from the sun (C in Figure). This causes the subsolar point to be as far south as it ever goes, 23°30' S (the Tropic of Capricorn). The circle of illumination is offset once again this time making the day short and the night long in the Northern Hemisphere. This is the Northern Hemisphere's Winter Solstice. Do you see that the rule regarding the location of the subsolar point holds. The subsolar point is as far south as it ever gets making the period the winter for the Northern Hemisphere. At the same time, this marks the beginning of the summer in the Southern Hemisphere. This event is technically called the December Solstice. Note once again where strange things happen. Figure shows that the Arctic Circle experiences 24 hours of darkness while the Antarctic Circle has 24 hours of daylight.

Climate Variability

This diagram shows the relationship between physical and biological oceanography and climate variability. Heat transport and ocean circulation are key factors between physical oceanography and climate variability. Biological oceanography impacts climate through the biological pump. Together, air-sea gas fluxes and penetrative solar radiation are feedbacks between physical and biological oceanography processes that ultimately influence climate.

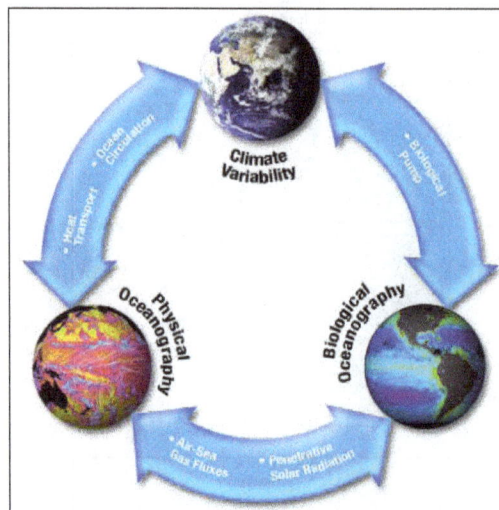

Climate is affected by both the biological and physical processes of the oceans. In addition, physical and biological processes affect each other creating a complex system. Both the ocean and the atmosphere transport roughly equal amounts of heat from Earth's equatorial regions - which are intensely heated by the Sun - toward the icy poles, which receive relatively little solar radiation. The atmosphere transports heat through a complex, worldwide pattern of winds; blowing across the sea surface, these winds drive corresponding patterns of ocean currents. But the ocean currents move more slowly than the winds, and have much higher heat storage capacity. The winds drive ocean circulation transporting warm water to the poles along the sea surface. As the water flows poleward, it releases heat into the atmosphere. In the far North Atlantic, some water sinks to the ocean floor. This water is eventually brought to the surface in many regions by mixing in the ocean, completing the oceanic conveyor belt. Changes in the distribution of heat within the belt are measured on time scales from tens to hundreds of years. While variations close to the ocean surface may induce relatively short-term climate changes, long-term changes in the deep ocean may not be detected for many generations. The ocean is the thermal memory of the climate system.

- Physical characteristics of heat transport and ocean circulation impact the Earth's climate system. Like a massive 'flywheel' that stabilizes the speed of an engine, the vast amounts of heat in the oceans stabilizes the temperature of Earth. The heat capacity of the ocean is F Trenberth K. E., Smith L., Qian T., Dai.

- much greater than that of the atmosphere or the land. As a result, the ocean slowly warms in the summer, keeping air cool, and it slowly cools in winter, keeping the air warm. A coastal city like San Francisco has a small range of temperature throughout the year, but a mid-continental city like Fargo, ND has a very wide range of temperatures. The ocean carries substantial heat only to the sub-tropics. Poleward of the sub-tropics, the atmosphere carries most of the heat.

- Climate is also influenced by the "biological pump," a biological process in the ocean that impacts concentrations of carbon dioxide in the atmosphere. The oceanic biological productivity is both a source and sink of carbon dioxide, one of the greenhouse gases that control climate. The "biological pump" happens when phytoplankton convert carbon dioxide and nutrients into carbohydrates (reduced carbon). A little of this carbon sinks to the sea floor, where it is buried in the sediments. It stays buried for perhaps millions of years. Oil is just reduced carbon trapped in sediments from millions of years ago. Through photosynthesis, microscopic plants (phytoplankton) assimilate carbon dioxide and nutrients (e.g., nitrate, phosphate, and silicate) into organic carbon (carbohydrates and protein) and release oxygen.

- Carbon dioxide is also transferred through the air-sea interface. Deep water of the ocean can store carbon dioxide for centuries. Carbon dioxide dissolves in cold water at high latitudes, and is subducted with the water. It stays in the deeper ocean for years to centuries before the water is mixed back to the surface and warmed by the sun. The warm water releases carbon dioxide back to the atmosphere. Thus the conveyor belt described below carries carbon dioxide into the deep ocean. Some (but not all, or even a large part) of this water comes to the surface in the tropical Pacific perhaps 1000 years later, releasing carbon dioxide stored for that period. The physical temperature of the ocean helps regulate the amount of carbon dioxide is released or absorbed into the water. Cold water can dissolve more carbon dioxide than warm water. Temperature of ocean is also impacted the biological pump. Penetrative solar radiation warms the ocean surface causing more carbon dioxide to be released into the atmosphere. Oceanic processes of air-sea gas fluxes effect biological production and consequentially impacting climate. But as plant growth increases, the water gets cloudy and prevents the solar radiation from penetrating beneath the ocean surface.

Sea Winds

Scatterometers are used to measure vector winds. The SeaWinds scatterometer has provided scientists with the most detailed, continuous global view of ocean-surface winds to date, including the detailed structure of hurricanes, wide-driven circulation, and changes in the polar sea-ice masses. Scatterometer signals can penetrate through clouds and haze to measure conditions at the ocean surface, making them the only proven satellite instruments capable of measuring vector winds at sea level day and night, in nearly all weather conditions. Combined with data from Topex/Poseidon, Jason-1, and weather satellites, moorings and drifters, data from SeaWinds and its follow-on missions will be used to study long-term change. Earth's weather patterns such as El Niño, and the Northern Oscillation, which affect the hydrologic and bio-geochemical balance of the ocean-atmosphere system.

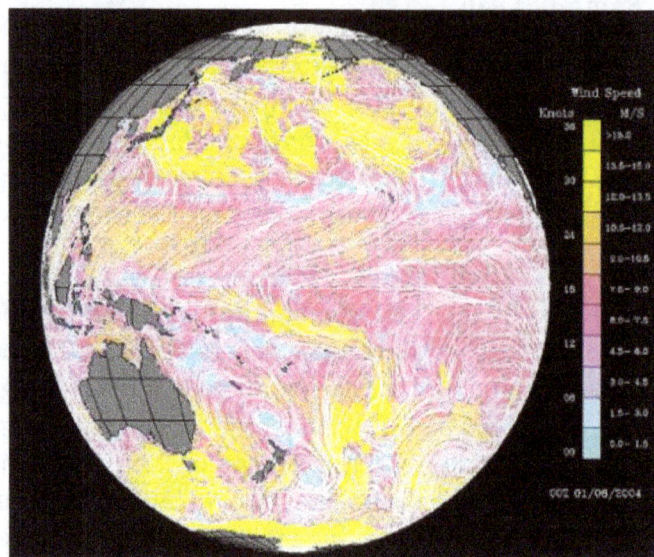

Ocean Surface Topography

Radar altimeters like those on the Topex/Poseidon and Jason missions, are used to measure ocean surface topography. Bouncing radio waves off the ocean surface and timing their return with incredible accuracy, these instruments tell us the distance from the satellite to the sea surface within a few centimeters - the equivalent of sensing the thickness of a dime from a jet flying at 35,000 feet! At the same time, special tracking systems on the satellites give their position relative to the center of mass of Earth also with an accuracy of a few centimeters. By subtracting the height of the satellite above the sea from the height of the satellite above the center of mass, scientists calculate maps of the sea-surface height and changes in the height due to tides, changing currents, heat stored in the ocean, and amount of water in the ocean. By mapping the topography of the ocean we can determine the speed and direction of ocean currents. Just as wind blows around high- and low-pressure centers in the atmosphere, water flows around the high and lows of the ocean surface.

TOPEX/Poseidon & Jason-1 View of Hurricane Isabel September 27, 2003. As Hurricane Isabel slammed into the North Carolina Coast this month, TOPEX/Poseidon and Jason-1 orbited calmly overhead. This is a false color illustration of wave height off the east coast of the United Stated on

September 15, 2003 shows a significant increase in wave height to over 5 meters beneath Hurricane Isabel.

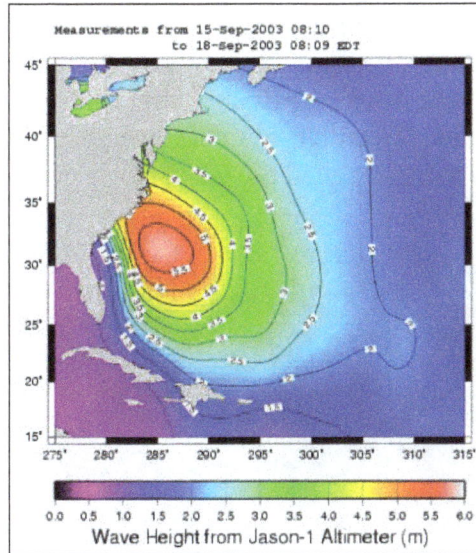

Maps of sea-surface height are most useful when they are converted to topographic maps. To determine topography of the sea-surface, height maps are compared with a gravitational reference map that shows the hills and valleys of a motionless ocean due to variations in the pull of gravity. The GRACE (Gravity Recovery and Climate Experiment) mission will provide very accurate maps of gravity that will allow us to greatly improve our knowledge of ocean circulation. GRACE has provided gravity measurements that are up to 100 times. This improved accuracy will lead the way to break-throughs in our understanding of ocean circulation and heat transport. Two figures showing sea surface height (SSH) and sea surface temperature (SST) Anomalies in the Pacific Ocean from October 1992 to August 2002. The increase in temperature and height in the equatorial region west of South America illustrates the 1997-98 El Nino event.

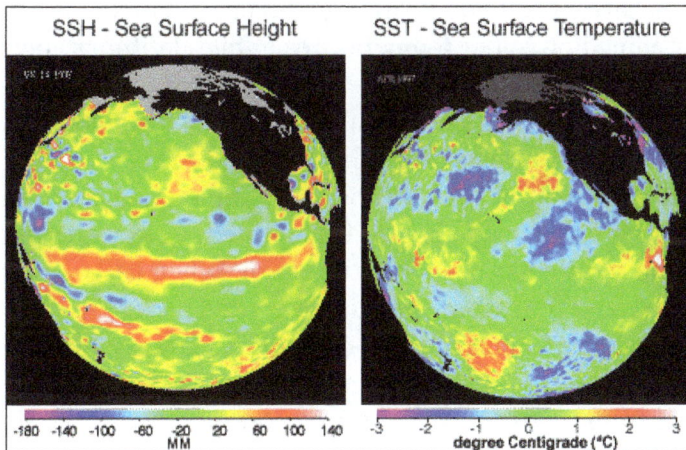

Sea-surface height is shown relative to normal with normal shown as green. Blue and purple areas represent heights measuring between 8 and 24 centimeters (3 and 9 inches) lower than normal. Red and white areas represent higher than normal sea-surface heights and indicate warmer water. These areas are between 8 and 24 centimeters, (3 and 9 inches) higher than normal.

Temperature & Salinity

Water is an enormously efficient heat-sink. Solar heat absorbed by bodies of water during the day, or in the summer, is released at night, or in winter. But the heat in the ocean is also circulating. Temperature & Salinity control the sinking of surface water to the deep ocean, which affects long-term climate change. Such sinking is also a principal mechanism by which the oceans store and transport heat and carbon dioxide. Together, temperature and salinity differences drive a global circulation within the ocean sometimes called the Global Conveyor Belt.

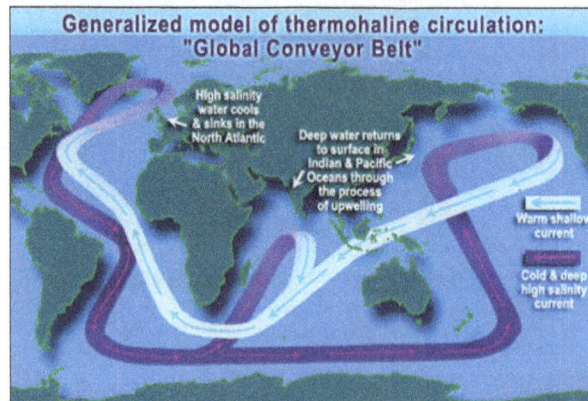

"The Global Conveyer Belt for Heat" represents in a simple way how ocean currents carry warm surface waters from the equator toward the poles and moderate global climate. This global circuit takes up to 1,000 years to complete. This illustration shows the generalized model of this thermo-haline circulation: 'Global Conveyor Belt.' Cold deep high salinity currents circulating from the north Atlantic Ocean to the southern Atlantic Ocean and east to the Indian Ocean. Deep water returns to the surface in the Indian and Pacific Oceans through the process of upwelling. The warm shallow current then returns west past the Indian Ocean, round South Africa and up to the North Atlantic where the water becomes saltier and colder and sinks starting the process all over again.

The heat in the water is carried to higher latitudes by ocean currents where it is released into the atmosphere. Water chilled by colder temperatures at high latitudes contracts (thus gets more dense). In some regions where the water is also very salty, such as the far North Atlantic, the water becomes dense enough to sink to the bottom. Mixing in the deep ocean due to winds and tides brings the cold water back to the surface everywhere around the ocean. Some reaches the surface via the global ocean water circulation conveyor belt to complete the cycle. During this circulation of cold and warm water, carbon dioxide is also transported. Cold water absorbs carbon dioxide from the atmosphere, and some sinks deep into the ocean. When deep water comes to the surface in the tropics, it is warmed, and the carbon dioxide is released back to the atmosphere. Salinity can be as important as temperature in determining density of seawater in some regions such as the western tropical Pacific and the far North Atlantic. Rain reduces the salinity, especially in regions of very heavy rain. Some tropical areas get 3,000 to 5,000 millimeters of rain each year. Evaporation increases salinity because as evaporation occurs, salt is left behind thus making surface water denser. Evaporation in the tropics averages 2,000 millimeters per year. This denser saltier water sinks into the ocean contributing to the global circulation patterns and mixing. Ocean salinity measurements have been few and infrequent, and in many places salinity has remained unmeasured. Remotely sensed salinity measurements hold the promise of greatly improving our ocean models. This is the challenge of project Aquarius, a NASA mission scheduled to launch in 2008, which will enable us to further refine our understanding of the ocean-climate connection.

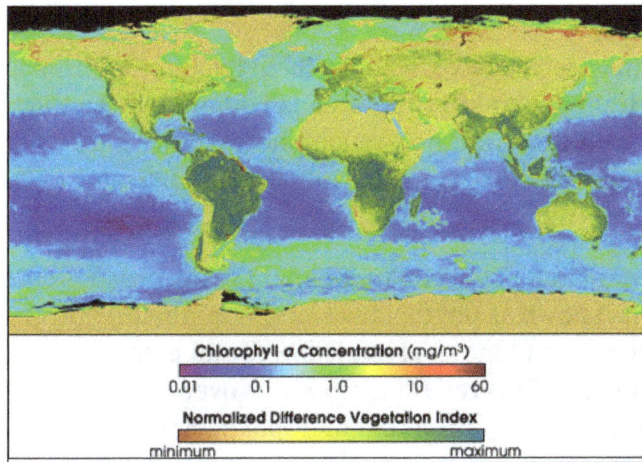

The above image shows the global biosphere. The Normalized Difference Vegetation Index (NDVI) measures the amount and health of plants on land, while chlorophyll a measurements indicate the amount of phytoplankton in the ocean. Land vegetation and phytoplankton both consume atmospheric carbon dioxide. This global biosphere image reveals amount of land vegetation in addition to amounts of phytoplankton. High amounts of phytoplankton are observed in the mid to high latitudes and along the west coast of North Africa and east coast of China.

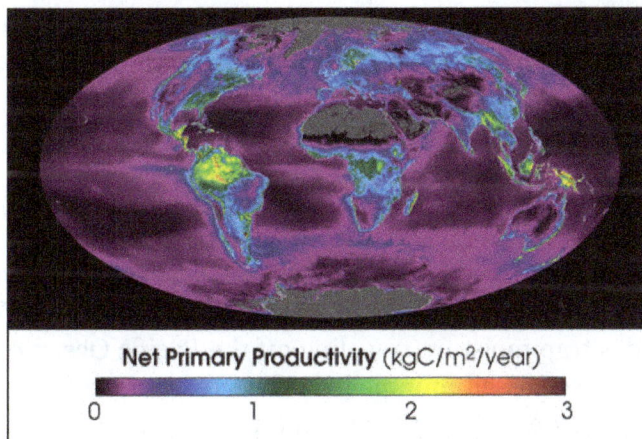

This false-color map represents the Earth's carbon "metabolism"-the rate at which plants absorbed carbon out of the atmosphere. The map shows the global, annual average of the net productivity of vegetation on land and in the ocean during 2002. The yellow and red areas show the highest rates, ranging from 2 to 3 kilograms of carbon taken in per square meter per year. The green, blue, and purple shades show progressively lower productivity.

The Biological Pump

Life in the ocean consumes and releases large quantities of carbon dioxide. Across Earth's oceans, tiny marine plants called phytoplankton use chlorophyll to capture sunlight during photosynthesis and use the energy to produce sugars. Phytoplankton are the basis of the ocean food web, and they play a significant role in Earth's climate, since they draw down carbon dioxide, a greenhouse gas, at the same rate as land plants. About half of the oxygen we breathe arises from photosynthesis in the ocean.

Because of their role in the ocean's biological productivity and their impact on climate, scientists want to know how much phytoplankton the oceans contain, where they are located, how their distribution is changing with time, and how much photosynthesis they perform. They gather this information by using satellites to observe chlorophyll as an indicator of the number, or biomass, of phytoplankton cells.

Probably the most important and predominant pigment in the ocean is chlorophyll-α contained in microscopic marine plants known as phytoplankton. Chlorophyll-α absorbs blue and red light and reflects green light. If the ratio of blue to green is low for an area of the ocean surface, then there is more phytoplankton present. This relationship works over a very wide range of concentrations, from less than 0.01 ton early 50 milligrams of chlorophyll per cubic meter of seawater.

Effects of Topography on the Climate

The effects of topography on the climate of any given region are powerful. Mountain ranges create barriers that alter wind and precipitation patterns. Topographical features such as narrow canyons channel and amplify winds. Mountains and plateaus are exposed to the cooler temperatures of higher altitudes. The orientation of mountains to the sun creates distinct microclimates in areas such as the Alps, where entire villages remain in the shade for most of the winter season.

Topography Affects Rain and Snowfall

Mountains play an important role in precipitation patterns. Topographic barriers such as mountains and hills force prevailing winds up and over their slopes. As air rises, it also cools. Cooler air is capable of holding less water vapor than warmer air. As air cools, this water vapor is forced to condense, depositing rain or snow on windward slopes. Mountains in the Western United States such as the Sierra Nevadas trap moisture traveling off the Pacific Ocean on their western flanks, where otherwise it might have passed unimpeded. This creates an effect known as a rainshadow on their leeward (protected) sides, where the air contains very little moisture. Most of the world's great mid-latitude deserts are located in rainshadows.

Topography Creates Distinctive Regional Winds

Mountain barriers also create and funnel regional winds, an important element of climate. As wind descends the leeward slopes, the air compresses, becoming more dense and warm. Strong winds can result, such as the powerful and unseasonably warm Chinook winds that flow down the eastern side of the Rocky Mountains. In arctic regions, extremely dense dry air is pulled off the edges of ice sheets by gravity. These forceful rushing winds are known as katabatic or gravity winds. Mountain passes also act as natural funnels and increase wind speeds. In California, Santa Ana winds blowing off the deserts are enhanced by these breaks. Wind blows more strongly when forced by topography through a narrow opening, and many wind farms can be found in these locations.

Higher Elevations and Cooler Temperatures

Land at higher elevations, such as mountains or plateaus, are naturally cooler due to a phenomenon known as the environmental lapse rate. First observed by the explorer and naturalist Alexander von Humboldt, air cools at 3.5 degrees Fahrenheit for every 1,000 feet of elevation gain. This is the equivalent of traveling hundreds of miles north, and creates a complex Highland climate with great diversity. In America's Southwest, deserts lie at the base of mountains that are topped with great Ponderosa pine forests because of the effects of elevation.

Orientation of Topography and Microclimates

The orientation of slopes in relation to the sun has a profound effect on climate. In the northern hemisphere, south-facing slopes are sunnier and support entirely different ecological communities than north-facing slopes. The south side of a mountain may experience spring conditions weeks or even months ahead of its north side. Where year-round snow or glaciers exist, they are nurtured by the shade provided by north- and west-facing slopes. In mountainous regions such as the Alps in Europe, entire villages may be cast in shade for months in winter, only to emerge again in the spring. In such communities, it is common to have a holiday to mark the reappearance of the sun.

Climate Modeling

Earth's Energy Budget

Figure shows observational estimates for Earth's global energy budget. Earth is heated by radiation from the sun and cooled by thermal radiation back to space. We will use this diagram to construct our first climate model below. Before we do this, a few more words on Earth's radiation balance at the top-of-the-atmosphere.

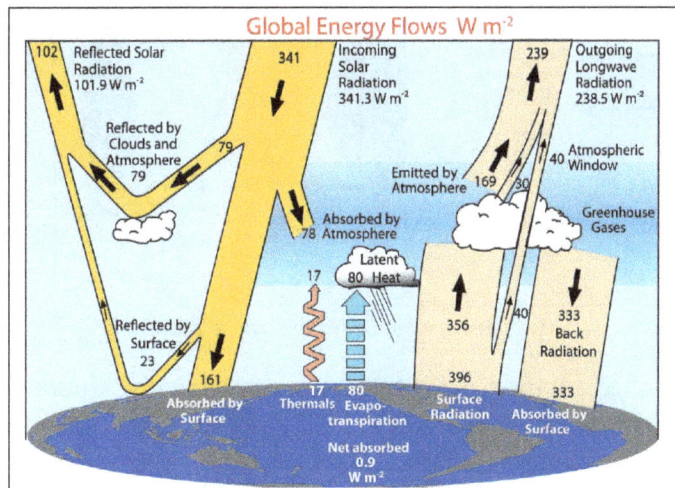

Earth's radiative budget.

The spectrum of energy emitted from a black body of a certain temperature T can be calculated from Planck's law of black body radiation. Figure shows the spectra for the Sun T_S = 5500 K and

Earth T_E = 255 K. The incoming solar radiation is at visible wavelength whereas the outgoing thermal radiation is in the infrared. The measured radiation is well approximated by a Planck curve. Some absorption occurs in the atmosphere, mainly due to water vapor and ozone, but most solar radiation penetrates to the surface.

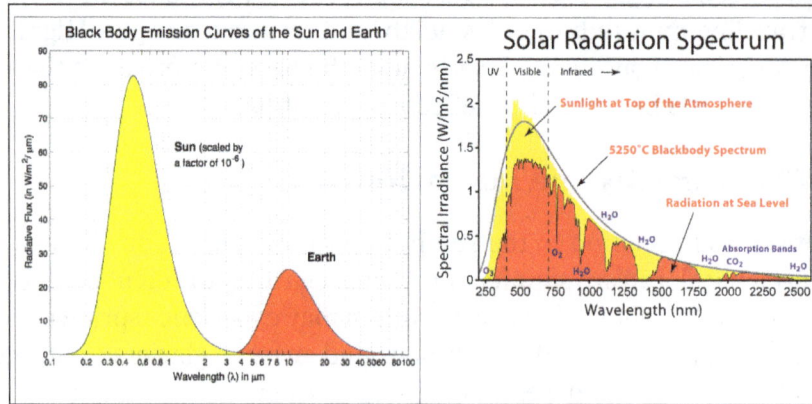

Left: Black body radiation for the Sun and the Earth.

The Zero-dimensional Energy Balance Model

The simplest model of the Earth's climate system is the zero-dimensional energy balance model (0D EBM). One-dimensional versions of these models were first developed by Budyko and Sellers. Earth is heated by solar radiation and cooled by radiating thermal (longwave) radiation back to space. The incoming shortwave radiation from the sun averaged over Earth's surface is S = S_o/4 = 342 W/m². The "solar constant" S_o = 1370 W/m² is the radiative flux through a disk with the radius of the Earth R = 6300 km. The factor 4 represents the average of this flux over the Earth's spherical surface area which is $4\pi R^2$, whereas a disk of radius R has the area πR^2.

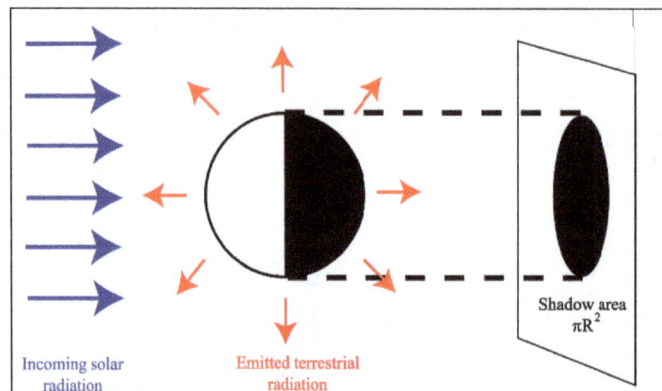

Heat absorbed and emitted by the Earth.

Some part of this incoming radiation is reflected back to space by clouds or snow and ice cover on the ground. This fraction is called planetary albedo and for Earth it is about a = 0.3. Assuming steady state, the shortwave radiation absorbed by the Earth's surface F_{sw} = (1-a) S must equal the longwave radiation $F_e = \sigma T_e^4$ emitted back to space at the equilibrium temperature T_e:

$$(1-a)S = \sigma T_e^4.$$

$\sigma = 5.67 \cdot 10^{-8}\,\text{W}/\,(\text{m}^2\text{K}^4)$ is the Stefan Boltzmann constant. Solving for T_e gives the table on the right. Thus, for Earth ($a = 0.3$) an equilibrium temperature of -18.2°C is predicted, which is more than 30°C colder than Earth's actual average surface temperature of about 15°C. The answer is that we didn't consider the atmosphere. In fact, Mars' temperature, who has no (or better a very thin) atmosphere, can be predicted reasonably well with equation $(1-a)S = \sigma T_e^4$.. Using a = 0.16 and $S^{\text{MARS}} = S^{\text{EARTH}}\,(1/1.52)^2$ in order to account for Mars' farther distance from the sun (1.52 times Earth's distance) gives an equilibrium temperature of -54.2°C, close to the observed value of -63°C.

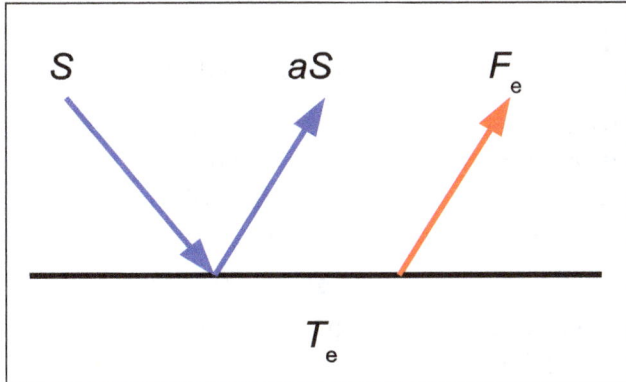

Surface radiative balance of a 0D EBM.

a	$T_e\,[°C]$
0.1	-1.7
0.3	-18.2
0.5	-38.2

However, gases in Earth's atmosphere, mainly water vapor and CO_2, act like the glass of a greenhouse such that they absorb much of the longwave radiation emitted from the surface. We can consider this by modifying our EBM as depicted in Fig. Now the atmosphere allows only a fraction τ of the surface radiation to be transmitted to space. It will assume a temperature Ta and this leads to emission of longwave radiation $F_a = \sigma T_a^4$ to both space and downward to the surface. Now we have two equations for the energy balance of the surface:

$$(1-a)S = F_e - F_a,$$

and that of the atmosphere

$$(1-\tau)F_e = 2F_a.$$

Inserting eq. $(1-\tau)F_e = 2F_a.$ into eq. $(1-a)S = F_e - F_a$, yields

$$(1-a)S = \frac{(1+\tau)}{2}F_e \equiv gF_e \equiv \tilde{F}_e.$$

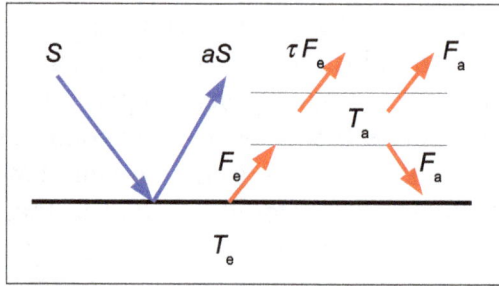

Radiative balance of a oD EBM including an atmosphere.

Note that eq. $(1-a)S = \dfrac{(1-\tau)}{2}F_e \equiv gF_e \equiv \tilde{F}_e.$ is very similar to eq. $(1-a)S = \sigma T_e^4.$ except for the greenhouse factor g. Choosing a value for Earth atmosphere's transmissivity of τ =0.23 and thus g = 0.62 a realistic surface temperature of T_e = 15°C is obtained. The atmosphere temperature of T_a = −46°C, which is close to the observed temperature of the tropopause. Emission of surface longwave radiation is Fe=390 W/m2, the total outgoing longwave flux at the top of the atmosphere $\tau F_e + F_a = 240 \text{ W} / \text{m}^2$, numbers in reasonable agreement with observational estimates shown in Fig., given that we neglect many processes such as surface sensible and latent heat fluxes or absorption of shortwave radiation by the atmosphere.

From eq. $(1-\tau)F_e = 2F_a.$ it follows that $\dfrac{F_e}{F_a} = \dfrac{2}{1-\tau}$ and hence $\dfrac{T_e}{T_a} = \left(\dfrac{2}{1-\tau}\right)^{1/4}$. Thus, for all values of τ the atmospheric temperature will always be lower than the surface temperature and it will go to zero as $\tau \Rightarrow 1$.

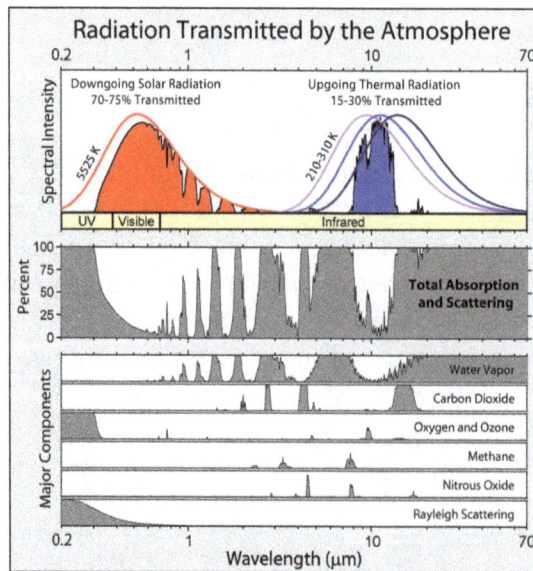

Atmospheric absorption spectra from a line-by-line modeling of molecular absorption using modern gas concentrations and assuming direct vertical transmission only.

Figure shows how the transmission of electromagnetic radiation through the atmosphere is influenced by absorption at different wavelengths from various atmospheric gases. Whereas the atmosphere is very transparent for shortwave radiation it is rather opaque for longwave radiation, similar to the properties of the glass in a greenhouse. (The glass in a greenhouse also has other effects such as inhibiting convection, which makes the analogy imperfect and the "greenhouse effect" perhaps even a misnomer.) The most important greenhouse gas is water vapor, but there is an important window around 10μ m in the water vapor absorption through which radiation from the surface can escape to space. However, the minor greenhouse gases CO_2, oxygen, ozone, methane and nitrous oxide absorb in this window, which makes them climatically important.

The Ice-Albedo Feedback

The longwave radiation increases as temperature increases. A small perturbation of the temperature from equilibrium, let's say a slight warming, will result in cooling through increased longwave radiation to space. Thus, the temperature dependence of the longwave radiation is a negative, or self-stabilizing, feedback mechanism. Another important feedback mechanism in the climate system is the ice-albedo feedback.

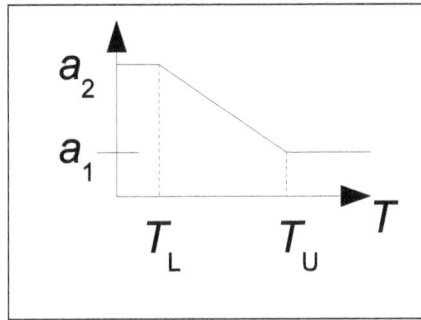

As temperatures drop below the freezing point snow and ice cover on the ground causes an increase of the surface albedo. Thus more sun light is reflected back to space and temperatures drop even further. This is a positive (self-amplifying) feedback.

In order to incorporate this feedback into our simple EBM we assume a ramp function for the planetary albedo:

$$a(T) = \begin{cases} a_1 = 0.3, T > T_U = 280K \\ a_1 + m(T_U - T), T_L \leq T < T_U \\ a_2 = 0.7, T < T_L = 250K \end{cases}$$

With $m = (a_2 - a_1)/(T_U - T_L)$ such that the albedo is low for climates warmer than an upper temperature TU and high if the climate is colder than T_L, with a linear transition in between.

We also linearize the longwave flux by developing $\tilde{F}_e = g\sigma T^4$ into a Taylor series around $T_0 = 288$ K:

$$\tilde{F}_e(T - T_0) = g\sigma T_0^4 + \frac{\partial \tilde{F}}{\partial T}\Big|_{T_0} (T - T_0) + ...,$$

$$\rightarrow \tilde{F}_e(T - T_0) = A + BT$$

With $A = -3g\sigma T_0^4 = -726W/m^2$ and $B = 4g\sigma T_0^3 = 3.36W/(m^2 K)$.

Equilibria of the system are now at:

$$\underbrace{(1-a(T))S}_{F_{SW}} = \underbrace{A+BT}_{F_{LW}},$$

and they can be found graphically from Fig. Three equilibria are possible. For our above set of parameters these are $T_1 = 14.3°C, T_2 = -12°C$ and $T_3 = -26°C$. The time dependent equation is:

$$C\frac{\partial T}{\partial t} = F_{SW} - F_{LW},$$

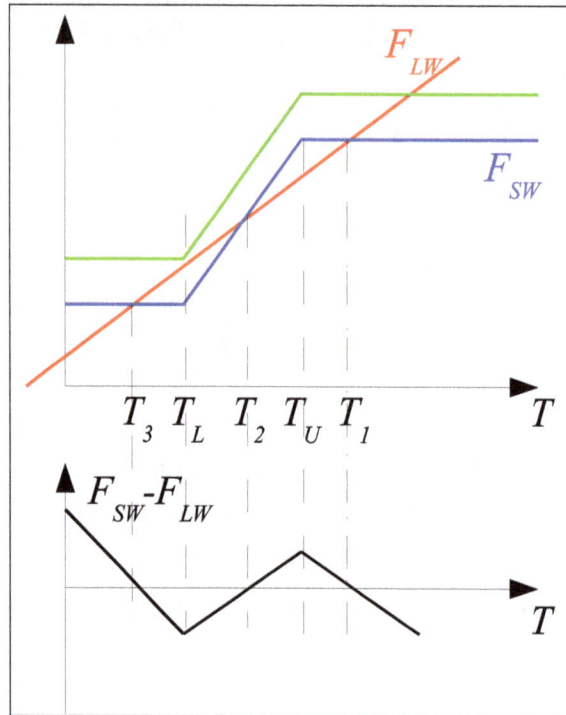

Three solutions of the 0D EBM with ice albedo feedback. The green line shows
a system with higher S_0 which exhibits only one equilibrium.

with the heat capacity C > 0. Thus equilibria with a negative slope of F_{SW}-F_{LW} are stable. E.g. a small positive perturbation to equilibrium T_1 will lead to an increase in outgoing longwave radiation whereas the shortwave stays constant. This leads to cooling and hence the perturbation will be damped. If the system is at T_2 a small positive perturbation leads to a larger increase in the shortwave radiation than the longwave, thus amplifying the perturbation. Hence T2 is an unstable equilibrium.

This can also be derived more formally by writing the total temperature $T = T_0 + T(t)$ as its value at the equilibrium T_0 plus a time dependent perturbation T'. Inserting into eq. $C\frac{\partial T}{\partial t} = F_{SW} - F_{LW}$, gives:

$$C\frac{\partial T'}{\partial t} = (1-a(T_0+T'))S - A - B(T_0+T')$$

At $T_0 = T_1$ we get:

$$C\frac{\partial T'}{\partial t} = (1-a_1)S - A - B(T_1 + T')$$

$$\rightarrow C\frac{\partial T'}{\partial t} = -BT'$$

The solution of the last partial differential equation is:

$$T' = T'(t=0)e^{-\frac{B}{C}t}.$$

The perturbation gets damped exponentially with the time scale C/B and thus the equilibrium is stable.

At $T_0 = T_2$ we get:

$$C\frac{\partial T'}{\partial t} = \underbrace{-(B-mS)}_{1.2}T',$$

an exponential growth of the perturbation. Thus, the equilibrium is unstable to small perturbations.

At $T_0 = T_3$ we get analogous to $T_0 = T_1$ a stable equilibrium. More generally, the stability of a time dependent system $\frac{\partial T}{\partial t} = f(T)$ at the equilibrium T_0 with $\frac{\partial T_0}{\partial t} = 0 = f(T_0)$ can be evaluated by assuming a small perturbation T' and linearizing $f\frac{\partial(T'_0 + T')}{\partial t} = f(T_0) + \frac{\partial f}{\partial T}\big|_{T_0} T'$ or $\frac{\partial T'}{\partial t} = \beta T'$, where $\beta = \frac{\partial f}{\partial T}(T_0)$ (Lyapunov exponent) with an exponential solution $T' = T'(t=0)e^{\beta t}$ The system is stable if $\beta < 0$, and unstable if $\beta > 0$.

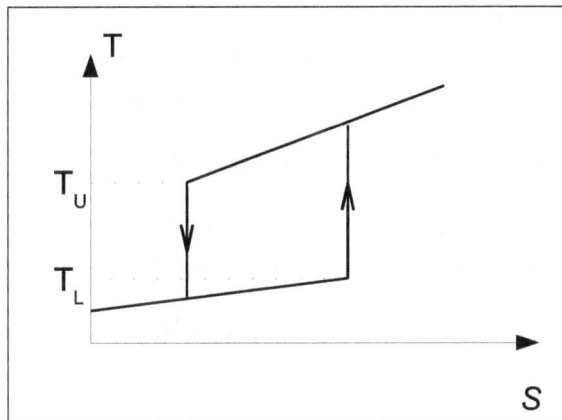

Hysteresis behavior of the 0D EBM with ice albedo feedback.

We conclude the 0D EBM with ice albedo feedback can have two stable equilibria. It is known from physics (e.g. magnetism) that a system with two stable equilibria can exhibit hysteresis behavior. That is, the state the system resides in at any given moment in time, does not only depend on the boundary conditions of the system (parameters of the model), but also on its history. Rapid transitions between the different equilibria can be triggered if a threshold of a slowly varying control parameter is passed.

These properties can be illustrated with Figs. by assuming the system resides in the cold state and S is slowly increased. This leads to $T_2 \to T_L$ and $T_3 \to T_L$. At $T_2 = T_3 = T_L$ the EBM switches to the warm state. Assume the solar constant is then decreased again. In order to return to the cold state it is not sufficient to lower S to the threshold $T_2 = T_3 = T_L$. The system will remain in the warm state until it be comes unstable at $T_2 = T_1 = T_U$, after which a rapid transition to the cold state occurs. Thus, the temperature difference $\Delta T = T_U - T_L$ determines the width of the hysteresis curve. If $\Delta T < \Delta T_{crit}$ at fixed $\Delta a = a_2 - a_1$ only one steady state is possible. Then,

$$m = -\frac{\Delta T}{\Delta a} < B = \frac{\partial F_{LW}}{\partial T}.$$

Now that we know the system from our analytical analysis let's program our first numerical climate model. In order to do so we need to discretize eq. $C\frac{\partial T}{\partial t} = F_{SW} - F_{LW}$. We replace the differentials by finite differences:

$$C = \frac{\Delta T}{\Delta t} = F_{SW} - F_{LW}$$

and use a constant time step $\Delta t = t_{n+1} - t_n$. The temperature change during one time step is $\Delta T = T_{n+1} - T_n$. Now we can calculate the new temperature at time index n + 1 from the previous temperature at time index n according to:

$$T_{n+1} = \frac{(F_{SW} - F_{LW})}{C}\Delta t + T_n.$$

This is called the "Euler forward" time differencing scheme. Now we only need to know the heat capacity C in order to write a program and run it forward in time. For the Earth's climate system $C = C_O + C_A = \rho_O C_{Op} H_O + \rho_A C_{Ap} H_A$ we take the sum of the ocean's C_O and the atmosphere's C_A heat capacity. The density of sea water is about $\rho_O = 1000 kg/m^3$, that of air $\rho_A = 1.2 kg/m^3$. The specific heat at constant pressure of sea water is about $C_{Op} = 4200 J/(kg\ K)$, that of air $C_{Ap} = 1000 J/(kg\ K)$, and the height of the atmosphere is $H_A = 8300m$ and for the depth of the ocean mixed layer we assume $H_O = 50m$. Thus the heat capacity of the ocean $C_O = 2.1 \cdot 10^8 J/(m^2 K)$ is about 20 times that of the atmosphere $C_A = 1 \cdot 10^7 J/(m^2 K)$ and the total heat capacity of the climate system is approximately $C = 2.2 \cdot 10^8 J/(m^2 K)$. From equation $T' = T'(t=0)e^{\frac{B}{C}t}$. we see that Earth's climate system damps perturbations with a characteristic timescale $C/B = 6.5 \cdot 10^7 s \approx 2a$ of about two years.

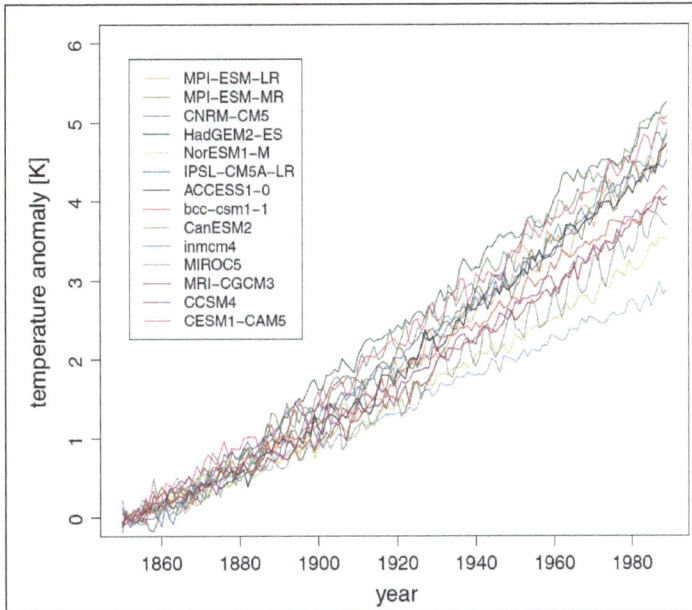

Evolution of global mean near surface air temperature, expressed as a difference to
a control run, due to an exponential (1%/yr) increase of atmospheric CO2.

Climate Sensitivity

Radiative Forcing is the instantaneous change of the radiative energy balance at the top of the
troposphere (after adjustment of the stratosphere) due to a change in something (e.g. greenhouse
gas concentrations, incoming solar radiation, surface albedo, aerosols) with everything else (e.g.
temperature, water vapor) fixed. Current state-of-the-science climate models show a range of response to a given forcing. Fig. illustrates this for an idealized numerical experiment in which CO_2
is increased by 1% per year from 280 ppm at model year 1850 to 1120 ppm. All 14 models respond
with warming but some models warm more than others such that at the end of the experiment the
temperature increase varies from ~ 3 K to ~ 5 K. The global surface air temperature change due
to a given forcing is referred to as the climate sensitivity; in this case, since climate is still changing,
the transient climate sensitivity due to a quadrupling of CO_2.

More commonly climate sensitivity is referred to as the global mean surface air temperature
increase for a doubling of atmospheric $CO_2 : \Delta T_{2xC}$. Its value is highly uncertain, which is an
important reason for the uncertainty in transient climate sensitivity illustrated by the range
in Fig. Narrowing this uncertainty remains a major challenge in current climate research. The
radiative forcing for a change of atmospheric CO_2 is relatively well known (to within 10%) and
depends on the logarithm of the CO_2 concentration C $\Delta Q = Q_0 \ln \left(C / C_0 \right)$, where C_0 is the
reference CO_2 concentration (e.g. the preindustrial value of 280 ppmv) and $Q_0 = 5.35$ W/m².
The logarithmic dependency is due to the near saturation of the main CO_2 absorption band.
Thus the radiative forcing for a doubling of CO_2 is $\Delta Q_{2xC} = 3.7$ W/m^2. An exponential increase
in CO_2 will therefore lead to a linear increase in radiative forcing, in case of Fig. the forcing
increases from zero to 7.4 W/m². The global mean temperature response of the models is also
approximately linear.

A more general definition of equilibrium climate sensitivity is the change in global mean surface air temperature ΔT for an arbitrary radiative forcing ΔQ after the climate system has reached a new steady state:

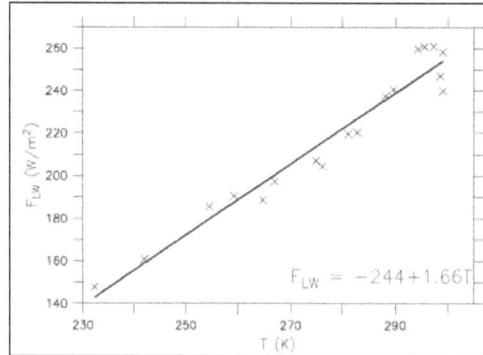

Figure: Regression of outgoing longwave radiation at the top-of-the-atmosphere (from ERBE satellite observations) versus surface air temperature (from NCEP re-analysis). All data have been averaged on 10 degree latitude bands.

$$\alpha = \frac{\Delta T}{\Delta Q}.$$

ΔQ can be either a change in shortwave or longwave fluxes. For our EBM in equilibrium (eq. $\underbrace{(1-a(T))S}_{F_{SW}} = \underbrace{A+BT}_{F_{LW}},)$ assuming a constant albedo we have

$$(1-a)S = A + BT_0 \text{ and}$$

$$(1-a)S + \Delta Q = A + B(T_0 + \Delta T).$$

Thus the climate sensitivity of our EBM is simply

$$\alpha = \frac{1}{B},$$

which gives $\alpha = 0.3K\left(Wm^{-2}\right)^{-1}$ for $B = 3.36\ Wm^{-2}K^{-1}$, which corresponds to a warming of $\Delta T_{2xC} = 1$ K for a doubling of CO_2. This is the climate sensitivity in the absence of feedbacks

GCMs (and most likely also the real climate system) have a larger climate sensitivity of about 0.6-1.1 K(Wm^{-2})$^{-1}$. What is the reason for this discrepancy? Above, at eq. $\rightarrow \tilde{F}_e(T-T_0) \dots A + BT$, when linearizing the longwave radiation we have been somewhat cursory and used the surface temperature $T_0 = T_e = 288$ K for the state to linearize around. However, most of the outgoing longwave radiation is emitted from the cold upper atmosphere. From eq. $(1-\tau)F_e = 2F_a$. it can be calculated that only 38% of the longwave radiation emitted to space comes from the surface. Thus it seems more appropriate to use the tropopause temperature $T_a = 227K$ as the state to linearize eq. $\rightarrow \tilde{F}_e(T-T_0) \dots A + BT$ around. If this is done we get A = -246 W/m^2 and B =1.67 W/(m^{2}K). These values for A and B are almost identical to the values one obtains by a regression of observed (by satellite measurements) longwave fluxes against surface air temperatures. Note that the latter determination of the parameters implicitly includes feedbacks such as the water vapor feedback. Using this value for B to recalculate the climate sensitivity we get a more reasonable value of $\alpha = 0.6K\left(Wm^{-2}\right)^{-1}$ or $\Delta T_{2xC} = 2.3K$, which is within the 66% probability range of 2-4.5 K reported by the IPCC. Most complex models fall within this range, but larger values cannot be excluded at present. Changes in albedo would increase the climate sensitivity. Using the oD EBM this contribution is difficult to estimate, because it neglects meridional differences. In the real world we would expect albedo changes due to changes in snow or ice cover to occur only at those latitudes that experience seasonal temperature variations including the freezing point. In other words, a 1K temperature change would not affect snow and ice cover in the tropics (because it is too warm) or over Antarctica (because it is too cold). Thus, albedo changes will strongly depend on latitude and we will use the one-dimensional version of the EBM below to estimate this contribution to the climate sensitivity.

A more general formulation of the climate sensitivity can be derived by assuming a new steady state with an additional forcing term ΔQ in the radiative balance at the top of the atmosphere. Let's assume $T_0 + \Delta T$ is the new surface temperature and T_o was the original equilibrium temperature. We also assume that the fluxes that change the surface temperature,

$$\frac{\partial T}{\partial t} = F\left(T, y_1, y_2, ..., y_n\right)$$

depend on T and n different variables y, at equilibrium $F_0 = F\left(T_0 y_{10}, ..., y_{n0}\right) = 0$. A radiative forcing can cause each parameter to change. Thus, for small changes in radiative forcing and temperature at the new equilibrium we get

$$\frac{\partial\left(T_0 + \Delta T\right)}{\partial t} = F_0 + \frac{\partial F}{\partial T}\Delta T + \Sigma_n\left(\frac{\partial F}{\partial y_n}\frac{\partial y_n}{\partial T}\right)\Delta T + \Delta Q = 0,$$

where the partial derivatives are taken at all other variables fixed except T (second term) or y_n (third term). The inverse of the climate sensitivity $\lambda = \alpha^{-1}$ is called feedback parameter. The total feedback parameter thus becomes

$$\lambda = \alpha^{-1} = \frac{\Delta Q}{\Delta T} = -\frac{\Delta F}{\Delta T} - \Sigma_n\left(\frac{\partial F}{\partial y_n}\frac{\partial y_n}{\partial T}\right) := \lambda_0 + \Sigma_n\lambda_n$$

the sum of the individual contributions from the different variables. Since the partial derivatives depend on T_o and the y_{no} it follows that the climate sensitivity depends on the background state. This can be illustrated with the surface (ice) albedo feedback. Consider a very warm climate state in which surface temperatures are above the freezing point everywhere. In this case the ice albedo feedback would be zero because the snow/ice cover would not change. For colder background states for which certain regions become snow covered the ice albedo feedback will be positive. It will get stronger the colder the climate gets since the area of snow cover will increase. However once the entire Earth is snow covered (Snowball Earth) the ice albedo feedback will once again be zero since small temperature changes will not affect snow cover.

The first term on the rhs of eq. $\lambda = \alpha^{-1} = \frac{\Delta Q}{\Delta T} = -\frac{\Delta F}{\Delta T} - \Sigma_n\left(\frac{\partial F}{\partial y_n}\frac{\partial y_n}{\partial T}\right) := \lambda_0 + \Sigma_n\lambda_n$ is the Planck feedback λ_0, which assumes a constant temperature change throughout the troposphere with all other variables fixed. Its value of $\lambda_0 = -3.23 \pm 0.03$ has a low uncertainty and is almost identical to the "no feedback" value we derived from the EBM in equation $\alpha = \dfrac{1}{B}$. Variables that have been shown to impact climate sensitivity are water vapor λ w, the lapse rate (change of temperature with height) λ_L, clouds λ_C, and surface albedo λ_a (e.g. through changes in vegetation or snow/ice cover). The total feedback parameter is the sum of these individual processes $\lambda = \lambda_0 + \lambda_w + \lambda_L + \lambda_C + \lambda_a$.

The main reason for the large range of climate sensitivities in coupled ocean-atmosphere models is their different cloud feedback λ_C. Clouds affect not only the albedo but longwave fluxes as well. The transient climate sensitivity, which is the temperature change at the time of a certain change in CO_2, is different from (and lower than) the equilibrium climate sensitivity mainly due to the large heat capacity of the oceans. However, models with higher equilibrium climate sensitivity usually also have higher transient climate sensitivities.

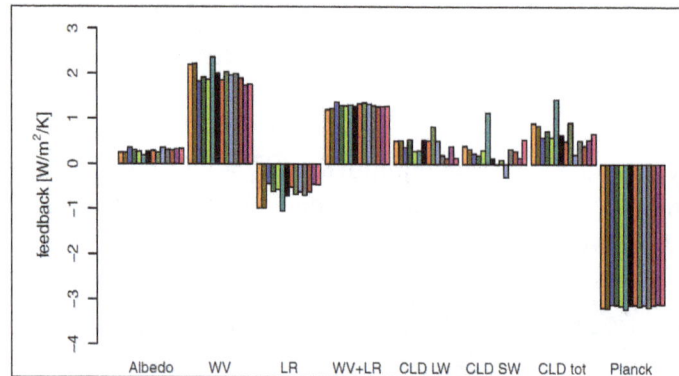

Figure: Analysis of feedback mechanisms in climate models that contribute to the climate sensitivity. Each bar represents a feedback parameter from a different model. Note that the water vapor and lapse rate feedbacks are related resulting in a smaller spread of their sum than the spread of the individual feedbacks. Cloud feedbacks, separated into effects on shortwave and longwave fluxes, show the large differences between models.

The water vapor feedback is due to an increase in water vapor in a warmer atmosphere according to the Clausius-Clapeyron equation. Because water vapor is a strong greenhouse gas this is a strong positive feedback effect.

The lapse rate feedback is due to the decrease in the moist adiabatic lapse rate with temperature. The atmospheric temperature profile in the tropics is close to the moist adiabat due to latent heat release. Thus, as the climate is warmed evaporation at the surface increases exponentially following the Clausius-Clapeyron equation. Increased evaporation tends to cool the surface. Hence the latent heat released at higher altitudes leads to a larger warming in the upper troposphere than at the surface. Following the 20°C moist adiabat from the surface to 8 km height in Figure you'll find a temperature of about -30°C. Now follow 25°C moist adiabat, which will lead to only -15°C at 8 km height. Thus, a 5°C warming at the surface becomes a 15°C warming in the upper troposphere due to the release of latent heat. Larger warming aloft tends to decrease the greenhouse effect. (Remember: the greenhouse effect is due to emissions to space at lower temperatures than the surface temperature. The definition of the lapse rate feedback is the feedback due to changes in the lapse rate without changes in the mean temperature. Thus, if the upper atmosphere warms more the surface must warm less. The lapse rate feedback is therefore negative.

Both water vapor and lapse rate feedback thus depend strongly on the hydrological cycle and vertical water vapor distributions. Thus it may not be surprising that the combined water vapor plus lapse rate feedback is less uncertain than the feedbacks individually.

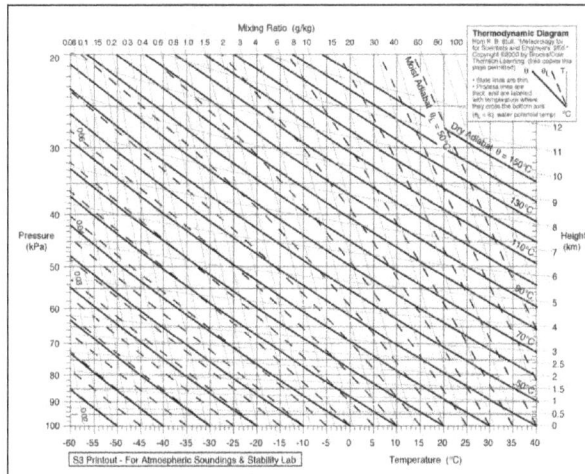

Dry (solid) and moist (dashed) adiabatic lapse rates as a function of surface temperature.

This concludes our analysis of the oD EBM. Although conceptually interesting it is of limited practical applicability owing to its simplicity and the neglect of spatial variability. In the following we will extend the EBM to one dimension including meridional heat transport and temperature variations with latitude.

Stochastic Climate Models

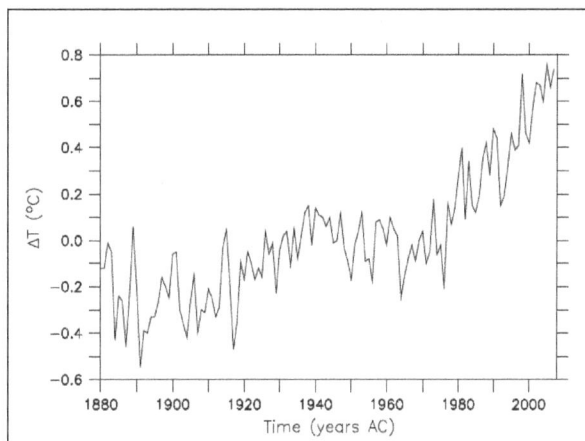

Global average surface air temperature anomaly from GISS.

Time series of real climate data are never smooth like the time series of our EBM but they contain a lot of variability on different time scales. A general property of real climate data is that their spectrum is red. This means that more variance is located at low frequencies than at high frequencies. Hasselmann suggested that this can be explained by the integrative nature of the climate system.

High frequency random variations (weather) are integrated by subsystems with a large heat capacity, such as the oceans or ice sheets, resulting in a red spectrum response of the climate system to the white noise forcing. On the other hand, in the real climate system there is forcing on long time scales due to slow changes of Earth's orbit around the sun. This generates variance at low frequencies.

An autoregressive process of order one (AR1) is a simple model in which the state at time $n + 1$ is determined by the state at the previous time n times a constant ($0 < b < 1$) plus a white noise term (w). Such a model produces a red spectrum and describes many observed climate time series to first order.

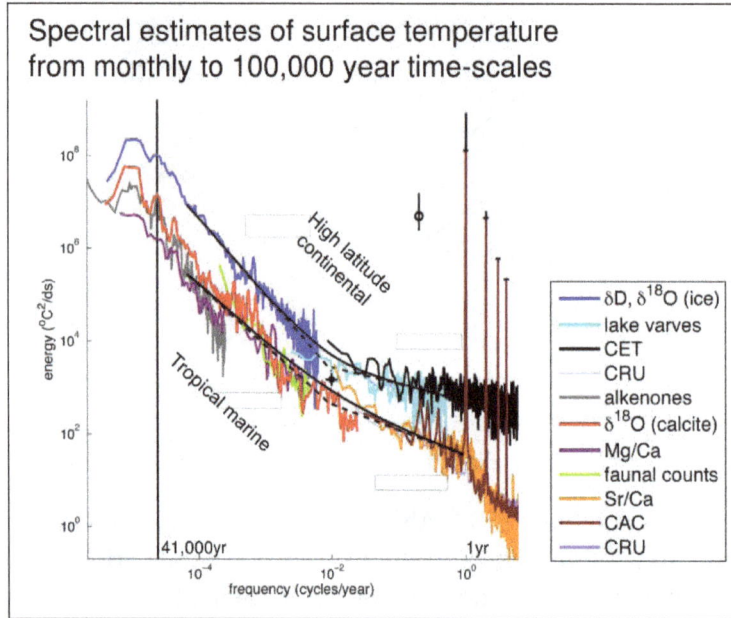

$$x_{n+1} = bx_n + w,$$

The One-dimensional Energy Balance Model

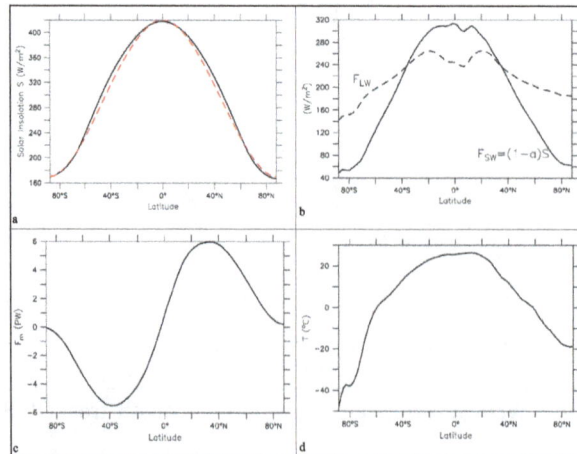

Figure: Zonally averaged radiative fluxes at the top-of-the-atmosphere from the Earth Radiation Budget Experiment (ERBE) as a function of latitude. (a) Solar Insolation S (black) and a simple analytical approximation $S(\varphi) = 295 + 125 \cos(2\varphi)$ (red), which we'll use for our 1D EBM. (b) Absorbed shortwave radiation (solid) and outgoing longwave radiation (dashed). At low latitudes the Earth receives more energy than it emits back to space, whereas at high latitude it looses more heat by longwave radiation than it receives from the sun. Integrating the net radiation from one pole yields the meridional heat transport by the climate system from low to high latitudes as plotted in (c) in PW (10^{15}W). (d) Zonally averaged surface air temperature from the NCEP Reanalysis.

Incident solar insolation S varies strongly with latitude as shown in figure. for the calculation of daily insolation as a function of Earth's orbital parameters. Together with higher albedo at higher latitudes this leads to a strong difference, of 240 W/m² or a factor of 4, of the absorbed solar radiation between the equator and the poles. In contrast the equator-to-pole difference in outgoing longwave radiation is only 50% (or ~ 100 W / m^2). The net radiation at the top-ofthe-atmosphere, that is the difference between the absorbed solar insolation and the outgoing longwave radiation F_{SW}–F_{LW}, is positive at low latitudes and negative a high latitudes, implying a meridional heat transport.

Meridional heat transport can be taken into account by using a diffusive parametrization:

$$\vec{F}_m = -CK\vec{\nabla}T = -CK\frac{\partial T}{\partial y},$$

with an eddy diffusivity K and y denoting the north-south direction in cartesian coordinates. This parametrization of the effect of transient eddies is appropriate at mid latitudes, for the largest spatial scales and for time scales longer than 6 months. We can calculate K as a function of latitude using the meridional heat flux from the satellite measurements and temperatures from the reanalysis. Large fluctuations of K at low latitudes (including an unphysical negative value at 5°N), where the mean circulation (Hadley cell) dominates the meridional heat transport, indicates that the diffusive parameterisation is problematic there. The equation for the temperature change at each latitudinal band becomes

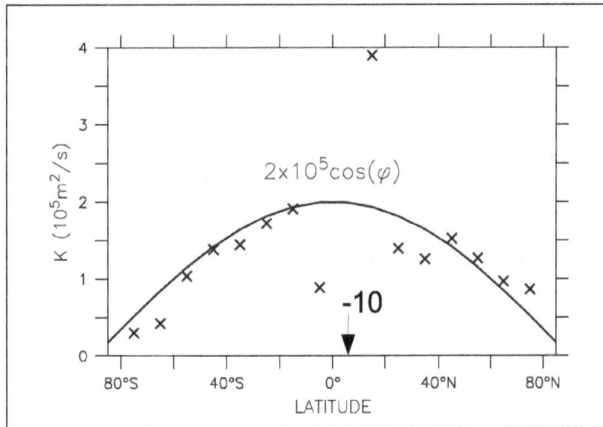

Meridional eddy diffusivity for heat (symbols) calculated from the observed meridional heat flux and temperature gradients displayed in Figure using equation $\vec{F}_m = -CK\vec{\nabla}T = -CK\frac{\partial T}{\partial y}$, . At 5°N a negative unphysical value of −10 was calculated. For our model we can use the simple analytical approximation shown as the solid line.

The equation for the temperature change at each latitudinal band becomes

$$C\frac{\partial T}{\partial t} = -\vec{\nabla}\vec{F}_m + F_{SW} - F_{LW},$$

Where the first term on the right hand side is the divergence of the meridional heat flux. All variables and parameters (e.g. $T(\phi)$, $C(\phi)$, $K(\phi)$) can now depend on latitude ϕ ranging from $-\pi/2$ at the south pole to $\pi/2$ at the north pole. Since Earth is a sphere we need to write the Laplace operator $\vec{\nabla}^2$ in spherical coordinates as

$$\vec{\nabla}\vec{F}_m = -\vec{\nabla}\left(CK\vec{\nabla}T\right) = \frac{-1}{R^2\cos\phi}\frac{\partial}{\partial\phi}\left(CK\cos\phi\frac{\partial T}{\partial\phi}\right).$$

In order to discretize eq. $C\dfrac{\partial T}{\partial t} = -\vec{\nabla}\vec{F}_m + F_{SW} - F_{LW}$, we set up a grid from 90°S to 90°N with N grid cells and constant grid spacing $\Delta\phi = \phi_{j+1} - \phi_j = \pi/N$. We will use a staggered grid, that is, we compute fluxes between two grid cells on their boundaries, whereas the variables (temperature) are evaluated in the center of each grid box. This leads to two latitude grids, the centers $\phi_j = 0.5\left(\tilde{\phi}_j + \tilde{\phi}_{j+1}\right)$ and the boundaries $\tilde{\phi}_j$. The meridional transport divergence becomes

$$-\vec{\nabla}\,\vec{F}_m = \frac{-1}{R\cos\phi}\frac{\Delta F_m}{\Delta\phi} = \frac{-1}{R\cos\phi}\frac{F_{mj+1} - F_{mj}}{\tilde{\phi}_{j+1} - \tilde{\phi}_j},$$

and the fluxes are computed at the boundaries

$$F_{mj} = -CK_j\frac{\cos\tilde{\phi}_j}{R}\frac{T_j - T_{j-1}}{\phi_j - \phi_{j-1}}.$$

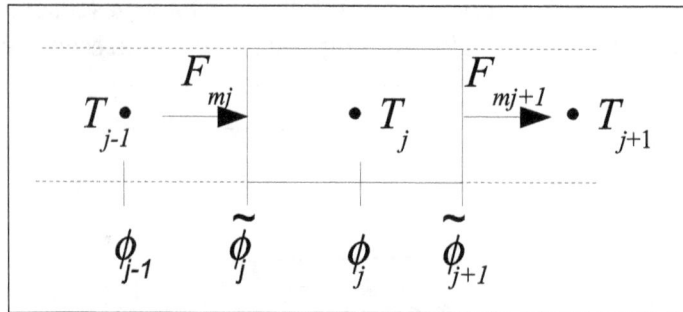

Staggered meridional model grid with temperatures T and fluxes F around grid point j.

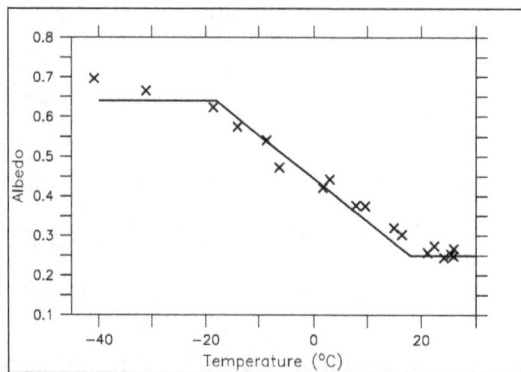

indicate, Albedo (from ERBE) as a function of surface air temperature (from NCEP) calculated from zonally averaged (on a 10° grid) data. The solid line shows a simple ramp function approximation

$$\text{(eq. } a(T) = \begin{cases} a_1 = 0.3, \, T > T_U = 280K \\ a_1 + m(T_U - T), T_L \le T < T_U \\ a_2 = 0.7, T < T_L = 250K \end{cases} \text{) with } T_L = -18°C, \, T_U = 18°C, \, a_1 = 0.64 \text{ and } a_2 = 0.25.$$

Numerics

If you increase the time step of your 1D EBM beyond a certain threshold the model will blow up. Using your EBM you can do another experiment using a relatively small value for the diffusivity. Determine the time-step threshold and use a value just below this threshold. This would be the most efficient time step to run your model. Now increase the diffusivity. You will notice that the critical time step depends on the diffusivity.

Numerics is an important issue in climate modeling. You need to make sure that your numerical solution is accurate and that it does not contain artifacts due to the way you discretize or time step the equations. There are different schemes to solve partial differential equations numerically and we want to investigate some simple examples below. Generally we want a scheme to have certain properties:

- Convergence for $\Delta x, \Delta t \to 0$,
- Stability,
- Accuracy,
- Conservation,
- Behavior of Amplitudes and Phases,
- Positive definite,
- No (or Small) Numerical Artifacts.

Generally a climate model is a numerical solution of a (set of) partial differential equation(s) as an initial value problem. The challenge will be to compute the interior points (empty circles in the graph below) from the initial conditions (black solid points) and the boundary conditions (grey points). This is also known as forward modeling.

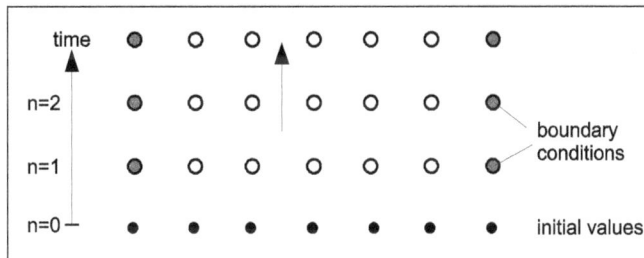

There are two types of boundary conditions:

- Dirichlet conditions specify values at the boundaries,
- Neuman conditions specify normal gradients at the boundaries.

Assume we know the solution T(t) at time t. Then we can develop a Taylor series:

$$T(t+\Delta t) = T(t) + \frac{dT}{dt}\Big|_t \Delta t - \frac{1}{2!}\frac{d^2T}{dt^2}\Big|_t (\Delta t)^2 + ...$$

such that,

$$\frac{dT}{dt}\Big|_t = \frac{T(t+\Delta t)-T(t)}{\Delta t} \underbrace{-\frac{1}{2!}\frac{d^2T}{dt^2}\Big|_t \Delta t - \frac{1}{3!}\frac{d^3T}{dt^3}\Big|_t (\Delta t)^2 - ...}_{correction\ of\ order\ \Delta t}$$

Neglecting terms of order Δt and higher we get the "Euler forward" scheme. The Euler scheme converges as $\Delta t \to 0$ to the true solution.

Now replace Δt with $-\Delta t$ in eq. and add this new equation to:

$$\frac{dT}{dt}\Big|_t = \frac{T(t+\Delta t)-T(t-\Delta t)}{2\cdot\Delta t} \underbrace{-\frac{1}{3!}\frac{d^3T}{dt^3}\Big|_t (\Delta t)^2 - .....}_{correction\ of\ order\ (\Delta t)^2}$$

This is the "centered differences" scheme. Corrections (errors) now scale with $(\Delta t)^2$ and approach zero faster than those of eq. $\frac{dT}{dt}\Big|_t = \frac{T(t+\Delta t)-T(t)}{\Delta t} \underbrace{-\frac{1}{2!}\frac{d^2T}{dt^2}\Big|_t \Delta t - \frac{1}{3!}\frac{d^3T}{dt^3}\Big|_t (\Delta t)^2 - ....}_{correction\ of\ order\ \Delta t}$

Consider as an example the centered differences scheme:

$$\frac{\partial C}{\partial x} \simeq \frac{C_{m+1}-C_{m-1}}{2\Delta x}$$

with a cosine wave $C = \hat{C}\cos(kx)$ represented numerically as $C_m = \hat{C}\cos(kx)$ and $C_{m+1} = \hat{C}\cos(k(x+\Delta x))$. We know that the exact solution is:

$$\frac{\partial C}{\partial x} = -\hat{C}k\sin(kx).$$

Using the trigonometric formula $\cos(x\pm y) = \cos(x)\cos(y) \mp \sin(x)\sin(y)$ we get for the numerical solution:

$$\frac{C_{m+1}-C_{m-1}}{2\Delta x} = \frac{\hat{C}}{\Delta x}\sin(kx)\sin(k\Delta x)\xrightarrow[\Delta x\to 0]{}\frac{\partial C}{\partial x}.$$

Thus the centered differences scheme converges against the true solution. The wave number can have values $k = 2\pi/(n\Delta x); n = 2, 3, ...$ such that the ratio between the numerical solution and the true solution becomes

$$\frac{C_{m+1}-C_{m-1}}{2\Delta x}\Big/\frac{\partial C}{\partial x} = \frac{\sin(k\Delta x)}{k\Delta x}$$

n	$\sin(k\Delta x)/(k\Delta x)$
3	0.41
4	0.64
6	0.82
8	0.9

thus, for fixed Δx only waves with large n (i.e. large wave lengths) are well represented. Waves with n < 8 have errors of more than 10%.

Numerical Solution of the Advection Equation

In our EBM we used a diffusion equation for the meridional transport of heat. In more complex fluid dynamical models advection equations are used for the transport of a property with the velocity of the fluid. Thus the advection equation is one of the most important equations in climate models and here we want to use it as an example to illustrate the properties of different numerical schemes. The advection equation in one dimension is

$$\frac{\partial C}{\partial t} = -\frac{\partial}{\partial x}(uC)$$

which describes the transport of property C with the fluid velocity u. Assuming a constant velocity everywhere we get

$$\frac{\partial C}{\partial t} = -u\frac{\partial C}{\partial x}.$$

An arbitrary function f is a solution of $\frac{\partial C}{\partial t} = -u\frac{\partial C}{\partial x}$. if

$$C(x,t) = f(x-ut).$$

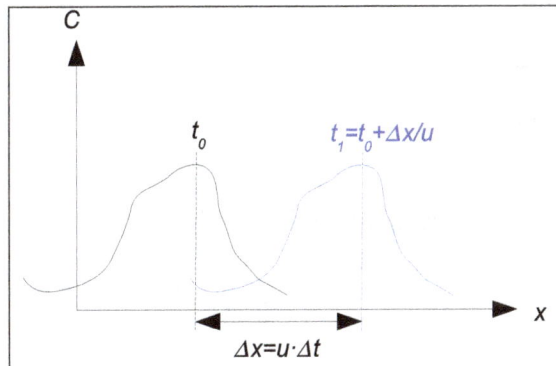

Von Neuman Stability Analysis

Now let's assume a wave function at time t = 0

$$C(x,0) = Ae^{ikx} = A(\cos(kx) + i\sin(kx)), \text{ with } i^2 = -1$$

Then at time t the solution is a plane wave:

$$C(x,t) = Ae^{ik(x-ut)}.$$

wave number	wavelength	angular frequency	period	frequency
$k = \dfrac{2\pi}{\lambda}$	$\lambda = \dfrac{2\pi}{k} = \dfrac{u}{v}$	$\omega = \dfrac{2\pi}{T}$	$T = \dfrac{2\pi}{\omega} = \dfrac{1}{v}$	$v = \dfrac{1}{T} = \dfrac{u}{\lambda}$

Now solve eq. $\dfrac{\partial C}{\partial t} = -u\dfrac{\partial C}{\partial x}$. numerically by discretizing time and space:

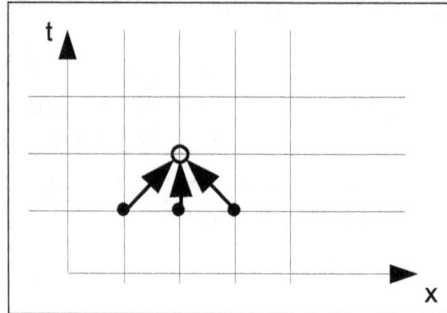

$$t = n\Delta t \qquad n = 0,1,2,....$$
$$x = m\Delta x \qquad m = 0,1,2,....$$

$$C(x,t) = C(m\cdot\Delta x, n\cdot\Delta t) = C_{m.n} = \xi^n e^{ikm\,\Delta x},$$

with the amplification factor $\xi(k)$. Each time step the solution is multiplied by ξ. Thus, if $\xi > 1$ the solution will diverge (blow up) and if it is $\xi < 1$ it will be damped.

Now let's examine the FTCS (forward in time centered in space) scheme:

$$\frac{C_{m,n+1} - C_{m,n}}{\Delta t} = -u\frac{C_{m+1,n} - C_{m-1,n}}{2\cdot\Delta x}$$

$$\rightarrow \xi = 1 - i\frac{u\Delta t}{\Delta x}\sin(k\Delta x)$$

Thus $|\xi| > 1$ for all k. The FTCS scheme is unconditionally unstable and therefore useless.

Now let's use centered differences $\dfrac{dT}{dt}\Big|_t = \dfrac{T(t+\Delta t) - T(t-\Delta t)}{2\cdot\Delta t} - \underbrace{\dfrac{1}{3!}\dfrac{d^3 T}{dt^3}\Big|_t (\Delta t)^2 .....}_{correction\ of\ order\ (\Delta t)^2}$ for

eq. $\dfrac{\partial C}{\partial t} = -\dfrac{\partial}{\partial x}(uC)$

$$\frac{C_{m,n+1} - C_{m,n-1}}{2\cdot\Delta t} = -u\frac{C_{m+1,n} - C_{m-1,n}}{2\cdot\Delta x}$$

$$C_{m,n+1} = C_{m,n-1} - \frac{u\cdot\Delta t}{\Delta x}\left(C_{m+1,n} - C_{m-1,n}\right)$$

This is the CTCS (centered in time, centered in space), or "leap-frog" scheme. The first time step has to be taken by a Euler scheme and two time steps in the past need to be stored in memory.

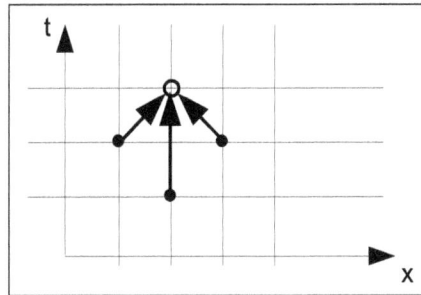

Now let's study the properties of the leap-frog scheme. Insert the analytical solution eq. $C(x,t) = C(m \cdot \Delta x, n \cdot \Delta t) = C_{m.n} = \xi^n e^{ikm \, \Delta x}$, in $C_{m,n+1} = C_{m,n-1} - \dfrac{u \cdot \Delta t}{\Delta x}\left(C_{m+1,n} - C_{m-1,n}\right)$:

$$\xi = \xi^{-1} - \frac{u\Delta t}{\Delta x}\left(e^{ik\Delta x} - e^{-ik\Delta x}\right)$$

$$\Leftrightarrow \xi^2 = 1 - 2i\,\sigma\,\xi$$

With $\sigma \dfrac{u\Delta t}{\Delta x}\sin(k\Delta x)$. The solution of this quadratic equation is:

$$\xi = -i\,\sigma \pm \sqrt{1-\sigma^2}$$

We distinguish two cases:

Instable case $|\sigma| > 1$:

$$\xi = -i(\sigma \pm S), \text{ with } S = \sqrt{\sigma^2 - 1} > 0.$$

If $\sigma > 1 \Rightarrow \sigma + S > 1 \Rightarrow |\xi^n| \to \infty.$

If $\sigma < -1 \Rightarrow \sigma - S < -1 \Rightarrow |\xi^n| \to \infty.$

Stable case $|\sigma| \leq 1$:

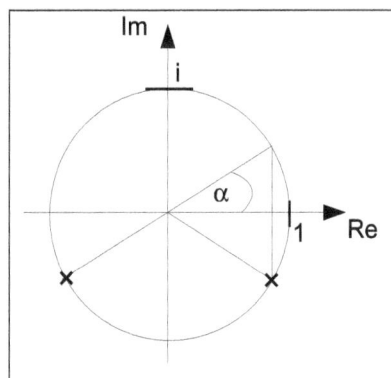

We can express sigma as a sine function $\sigma = \sin(\alpha)$ and using the trigonometric relation $\sin^2(\alpha) + \cos^2(\alpha) = 1$ we see that the solution of $\xi = -i\sin(\alpha)$ has an absolute value of one, it lies on the unit circle in the complex plane

$$\xi = \left\{ \begin{array}{c} e^{-i\alpha} \\ e^{i(\alpha+\pi)} \end{array} \right\}, \text{ with}$$

Now insert this in eq. $C(x,t) = C(m \cdot \Delta x, n \cdot \Delta t) = C_{m.n} = \xi^n e^{ikm\,\Delta x}$, we get

$$C_{m,n} = \left(M e^{-i\alpha n} + E e^{i(\alpha+\pi)n} \right) e^{ikm\Delta x}$$

And

$$C_{m,0} = (M + E) e^{ikm\Delta x},$$

thus with (2.29) $A = M + E$ or

$$C_{m,n} = \underbrace{(A-E)e^{ik\left(m\Delta x - \frac{\alpha n}{k}\right)}}_{P} + \underbrace{(-1)^n \, E e^{ik\left(m\Delta x + \frac{\alpha n}{k}\right)}}_{N},$$

with a physical mode P, and a numerical mode N, which changes sign each time step. Now we only have to determine E. For the first time step we have

$$C_{m,1} = C_{m,0} - \frac{u\Delta t}{2\Delta x}\left(C_{m+1,0} - C_{m-1,0}\right)$$

with $C_{m,0} = (M + E) e^{ikm\Delta x}$, we get

$$C_{m,1} = A\left(1 - i\sin(\alpha)\right)e^{ikm\Delta x} = (A-E)e^{ikm\Delta x - i\alpha} - E e^{ikm\Delta x + i\alpha}$$

Solve for E and enter into eq. $C_{m,n} = \underbrace{(A-E)e^{ik\left(m\Delta x - \frac{\alpha n}{k}\right)}}_{P} + \underbrace{(-1)^n \, E e^{ik\left(m\Delta x + \frac{\alpha n}{k}\right)}}_{N}$, yields

$$C_{m,n} = A\frac{1+\cos(\alpha)}{2\cos(\alpha)}e^{ik\left(m\Delta x - \frac{\alpha n}{k}\right)} + (-1)^n \, A\frac{1-\cos(\alpha)}{2\cos(\alpha)}e^{ik\left(m\Delta x + \frac{\alpha n}{k}\right)}.$$

It can be shown that $C_{m,n} = A\dfrac{1+\cos(\alpha)}{2\cos(\alpha)}e^{ik\left(m\Delta x - \frac{\alpha n}{k}\right)} + (-1)^n \, A\dfrac{1-\cos(\alpha)}{2\cos(\alpha)}e^{ik\left(m\Delta x + \frac{\alpha n}{k}\right)}$. converges to

$C(x,t) = A e^{ik(x-ut)}$ provided $\Delta x \to 0$ it follows that $\sigma \to uk\Delta t$ and for $\Delta t \to 0$ it follows that $\sigma \ll 1$ and hence $\sigma \sin(\alpha) \simeq \alpha$ and (2.40) converges to

$$C_{m,n} \to \underbrace{A\frac{1+\cos(\alpha)}{2\cos(\alpha)}e^{ik(x-ut)}}_{P} + \underbrace{(-1)^n \, A\frac{1-\cos\alpha}{2\cos(\alpha)}e^{ik(x+ut)}}_{N} \to A e^{k(x-ut)}$$

Thus, the leapfrog scheme is stable (provided $|\sigma| \leq 1$) and it converges against the true solution. However, for finite time steps and finite grid spacing a numerical solution N appears, which is unphysical. The physical solution P describes a plane wave traveling towards the right, whereas N changes sign every time step and travels towards the left.

The condition for stability $|\sigma| = |(u\Delta t / \Delta x)\sin(k\Delta x)| \leq 1$ must hold for all wavelength, thus it follows that $|(u\Delta t)/(\Delta x)| \leq 1$, which can be regarded as a condition for the maximum time step

$$\Delta t \leq \frac{\Delta x}{|u|}.$$

Equation $\Delta t \leq \dfrac{\Delta x}{|u|}$ is the CFL criterion. Physically, this criterion expresses the fact that the information flow in our numerical model is limited between neighboring grid cells. Thus, within one time step information can be transported maximally one grid cell. However, if the advection velocity is larger than $\Delta x / \Delta t$ information is transported farther than one grid cell. Therefore, in climate models the time step must always be smaller than the grid spacing divided by the maximum velocities. For a large scale ocean circulation model, for example, velocities of up to 1 m/s can occur. Given a grid spacing of 3° or 300 km the time step must be smaller than 3 days. An atmospheric model with the same resolution needs to use a time step of about 1 hour since the maximum velocities are much larger (~80 m/s in the jet stream) than those in the ocean.

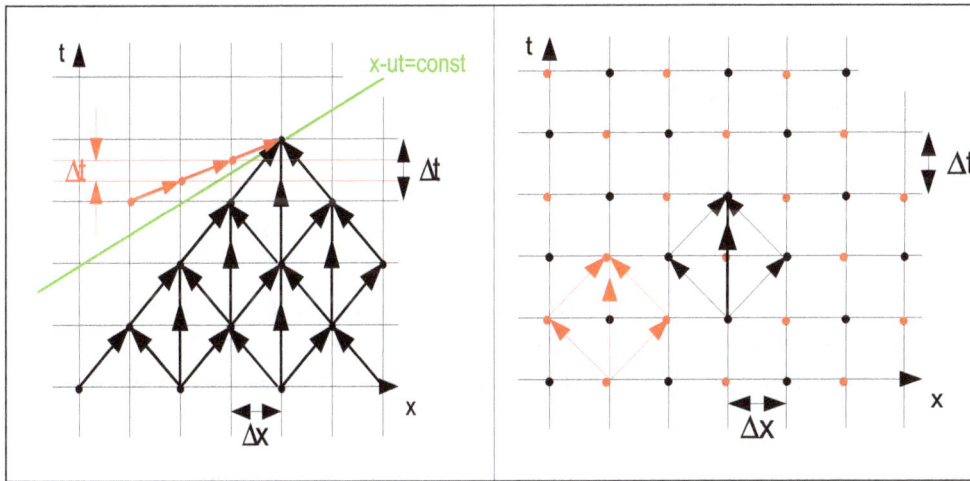

Figure: Left: Illustration of CFL criterion. If the time step is too large (black arrows) the information in the numerical model cannot propagate with the advection velocity (green line). Reducing the time step (red arrows) leads to a broadening of the cone of information allowing propagation of the information with the advection velocity. Right: Illustration of grid decoupling in leapfrog scheme. A chess board pattern of information transfer appears in which the red grid points do not communicate with the black grid points.

The numerical mode occurring in the leapfrog scheme is artificial. It is due to the decoupling of two grids as illustrated in the right panel of Figure. The numerical mode can be removed by using an Euler forward step (FTCS) once in a while.

The upwind scheme is illustrated below.

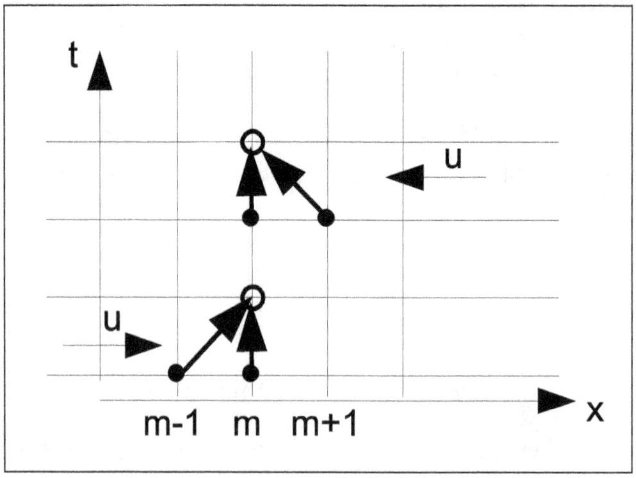

$$\frac{C_{m,n+1}-C_{m,n}}{\Delta t}=-u\left(\begin{array}{l}\dfrac{C_{m,n}-C_{m-1,n}}{\Delta x},u>0\\[4mm]\dfrac{C_{m+1,n}-C_{m,n}}{\Delta x},u\leq 0\end{array}\right)$$

For this scheme the amplification factor is

$$\xi=1-\left|\frac{u\,\Delta t}{\Delta x}\right|\big(1-\cos\left(k\Delta x\right)\big)-i\frac{u\,\Delta t}{\Delta x}\sin\left(k\,\Delta x\right)$$

$$\left|\xi^{2}\right|=1-2\left|\frac{u\,\Delta t}{\Delta x}\right|\left(1-\left|\frac{u\,\Delta t}{\Delta x}\right|\right)\big(1-\cos\left(k\,\Delta x\right)\big)$$

We find again the CFL criterion as a condition for stability. This scheme is good in a physical sense since properties are advected only from the direction of the velocity. However, the upwind scheme is only accurate to first order in the spatial derivatives and therefore it has "numerical diffusion".

Other properties of a numerical scheme might be important for particular applications. In some cases, e.g. for the simulation of atmospheric chemistry or ocean biology, it is important that the scheme is positive definite. This means that the concentration cannot become negative. Negative concentrations of chemical or biological species would lead to problems in the calculation of the source terms. Consider a chemical species such as nitrous oxide (N_2O) that is removed by photo-dissociation in the upper atmosphere. The removal rate is proportional to the concentration of the species. If small negative concentrations would occur due to errors in the numerical scheme this would lead to production of N_2O rather than destruction.

An example of a highly accurate, positive definite scheme without numerical diffusion is the Prather scheme. However, this scheme comes at the cost of much more required memory because it stores higher order moments of the tracer distributions.

Numerical Solution of the Diffusion Equation

Next we want to examine the diffusion equation, which we have already used in our 1D EBM

$$\frac{\partial C}{\partial t} = K \frac{\partial^2 C}{\partial x^2}.$$

The FTCS scheme leads to the following discretization:

$$\frac{C_{m,n+1} - C_{m,n}}{\Delta t} = K \frac{C_{m+1,n} - 2C_{m,n} + C_{m-1,n}}{\Delta x^2},$$

Or

$$C_{m,n+1} = C_{m,n} + \frac{K \Delta t}{\Delta x^2} \left(C_{m+1,n} - 2C_{m,n} + C_{m-1,n} \right).$$

This leads to

$$\xi = 1 - \frac{4K \Delta t}{(\Delta x)^2} \sin^2 \left(\frac{k \Delta x}{2} \right),$$

Or

$$\xi^2 = 1 - 2 \frac{4K \Delta t}{(\Delta x)^2} \sin^2 \left(\frac{k \Delta x}{2} \right) + \left(\frac{4K \Delta t}{(\Delta x)^2} \right)^2 \sin^2 \left(\frac{k \Delta x}{2} \right).$$

The condition for stability $|\xi| \leq 1$ implies that

$$\Delta t \leq \frac{(\Delta x)^2}{2K}.$$

This is the analog to the CFL criterion (eq. $\Delta t \leq \frac{\Delta x}{|u|}$.) for the diffusion equation. Now the minimum time step depends on the diffusivity. Surprisingly though, the FTCS scheme is stable for the diffusion equation, whereas it is unstable for the advection equation.

Implicit Schemes

Consider again the diffusive equation with the FTCS scheme and replace n on the right hand side of equation $\frac{C_{m,n+1} - C_{m,n}}{\Delta t} = K \frac{C_{m+1,n} - 2C_{m,n} + C_{m-1,n}}{\Delta x^2}$, with n + 1.

$$\frac{C_{m,m+1} - C_{m,n}}{\Delta t} = K \frac{C_{m+1,n+1} - 2C_{m,n+1} + C_{m-1,n+1}}{\Delta x^2}$$

This is the fully implicit or backward in time scheme. It can be solved by solving a set of linear equations

$$-\alpha C_{m-1,n+1} + (1+2\alpha) C_{m,n+1} - \alpha C_{m+1,n+1} = C_{m,n},$$

With $\alpha = K\Delta t / (\Delta x)^2$. This is a tridiagonal system that can be solved. The implicit scheme is unconditionally stable for any Δt! However, the implicit scheme is only of order one accurate and for the advection equation it displays numerical diffusion.

The Two-dimensional Energy Balance Model

Now add zonal resolution and transport:

$$\vec{\nabla}\vec{F} = -\vec{\nabla}\left(CK\vec{\nabla}T\right) = -\underbrace{\frac{1}{R^2\cos^2\phi}\frac{\partial}{\partial\lambda}\left(CK^\lambda(\phi)\frac{\partial T}{\partial\lambda}\right)}_{-F^\lambda} - \frac{1}{R^2\cos\phi}\frac{\partial}{\partial\phi}\left(CK^\phi(\phi)\cos\phi\frac{\partial T}{\partial\phi}\right)$$

assuming isotropic diffusion $K^\lambda = K^\phi = K(\phi)$. We can discretize the zonal heat flux and the heat flux divergence analogous to the treatment of the meridional fluxes

$$F_{i,j}^\lambda = -\frac{K_j}{R}\frac{T_{i,j} - T_{i,1j}}{\lambda_i - \lambda_{i-1}}$$

$$\vec{\nabla}F = -\frac{1}{R\cos^2\phi}\frac{F_{i+1,j}^\lambda - F_{i,j}^\lambda}{\Delta\lambda} - \dots$$

Use a zonal grid with cyclic boundary conditions FM + 1 = F1, TM + 1 = T1 and M = 36 boxes with a grid spacing $\Delta\lambda = 2\pi / M = 10°$

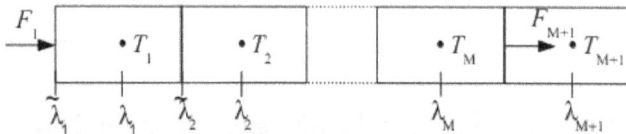

Test your 2D model first. The results should be zonally symmetric and the same as for the 1D model. Now introduce zonal asymmetry by constructing an idealized land sea mask and use spatially dependent heat capacity

$$C(\phi,\lambda) = \begin{cases} C_A, \ land \\ C_O = 20C_A, \ ocean \end{cases}$$

The Hydrological Cycle

Water cycles through all components of the climate system. Evaporation from the oceans and water vapor transport and precipitation over land are essential to provide freshwater to terrestrial ecosystems. Phase changes of water are also associated with release or absorption of latent heat. Evaporation leads to cooling of the surface and condensation in the atmosphere releases latent heat warming the surrounding air. This vertical transport of latent heat is important in determining the atmospheric lapse rate. Horizontal transport of water vapor and latent heat has important implications for temperature distributions and ecosystems.

The global hydrological cycle.

The hydrological cycle presents a particularly difficult challenge for climate models. Clouds, precipitation, permafrost, groundwater flow, and interactions between the ocean and ice shelfs are some of the major issues in climate modeling.

The Carbon Cycle

The global carbon cycle. Black (red) arrows and numbers represent pre-industrial (anthropogenic) fluxes.

Carbon is an essential element of all living organisms. It is taken up by plants during photosynthesis and released during respiration and oxidation of organic matter by bacteria both on land and in the ocean. On land most carbon is locked in living vegetation and in soils. Ocean biota on the other hand contain only a very small amount of carbon, most of which in microscopic plankton. Phytoplankton— plants that rely on photosynthesis for their growth—have to be small and light so that they don't sink out of the sun-lit upper ocean into the dark abyss. But they reproduce quickly and thus have similar rates of net primary production (carbon uptake) than the land biosphere. Some of the carbon fixed by ocean biota sinks into the deep ocean, where it is sequestered for a long time from the atmosphere. This process, called the biological pump, decreases CO_2 concentrations in the surface ocean and atmosphere. Thus, without ocean biology CO_2 concentrations in the atmosphere would be higher and climate would be warmer. In contrast to land, most carbon in the ocean occurs in inorganic form (dissolved inorganic carbon).

The ocean contains more than 40 times as much carbon than the atmosphere. Humans have increased atmospheric CO_2 concentrations by burning of fossil fuels and land use changes. Some of the excess carbon is taken up by the ocean and some is taken up by vegetation and soils on land.

The carbon cycle presents a challenge to modeling no less than the hydrological cycle. Biological systems are obviously complex. Many different species of plants and animals interact in intricate food webs and ecosystems. Microbes have important impacts on biogeochemical cycles such as those for nitrogen and oxygen. Since biological organisms require not only carbon but also other nutritional elements such as nitrogen, phosphorous, and iron, the cycles of those elements are important too.

Atmosphere

Radiative-convective Models

The simple energy balance model of Figure can be modified and extended to include more layers as shown in Figure. Now we assume the atmosphere is transparent to shortwave radiation and atmospheric layers 1 and 2 are completely opaque for longwave radiation. Further assuming that the atmospheric layers are perfect black bodies, the energy balance at the top of the atmosphere becomes

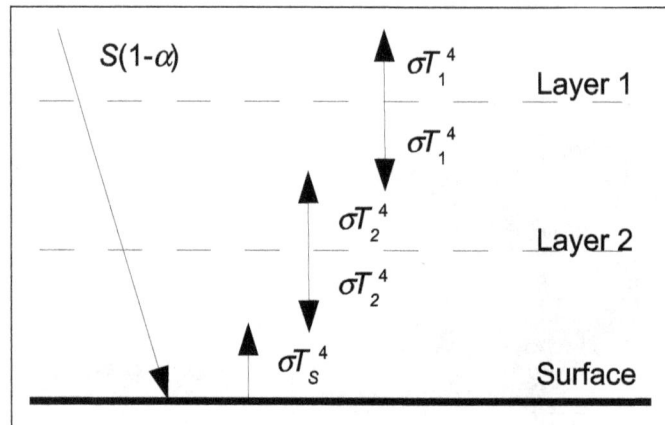

Simple two-layer radiative equilibrium model.

$$S(1-\alpha) = \sigma T_1^4.$$

For layer 1 the energy balance is

$$\sigma T_2^4 = 2\sigma T_1^4 = 2S(1-\alpha),$$

for layer 2 we have

$$\sigma T_1^4 + \sigma T_s^4 = 2\sigma T_2^4 = 4S(1-\alpha),$$

and at the surface

$$S(1-\alpha) + \sigma T_2^4 = \sigma T_s^4.$$

We notice that the temperatures increase downward. Solving these equations for the surface temperature we get

$$T_s^4 = 3S\frac{1-\alpha}{\sigma} = 3T_1^4.$$

Extending the model to n layers we see that the surface temperature is equilibrium will be always be larger than the temperature of the upper layer.

$$T_s = \sqrt[4]{n+1}\,T_1.$$

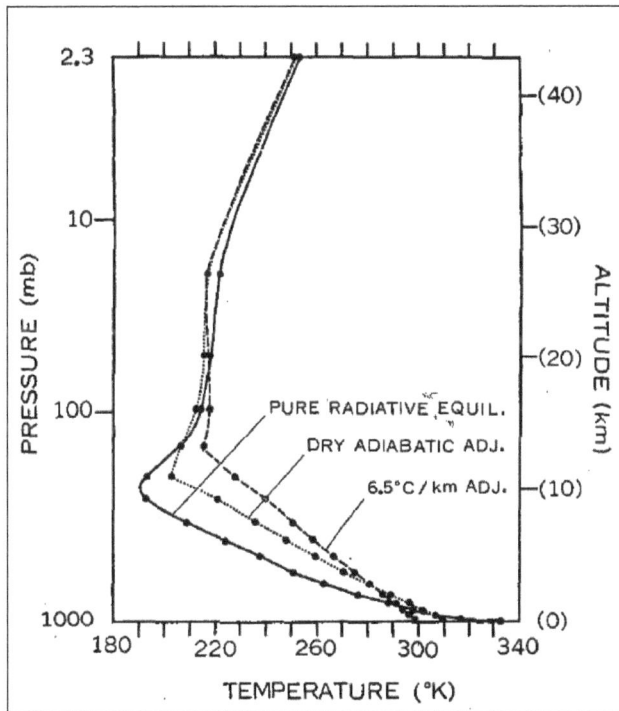

Vertical equilibrium temperature distribution calculated with a pure radiative transfer model (solid) and a radiative convective model using different maximum lapse rates.

For 2 layers the surface temperature is T_s = 335 K and the atmospheric temperatures are T_2 = 303 K and T_1 = 255 K. We see that the surface temperature is much too warm compared to the observed surface temperature of the Earth of about 288 K. What could be the reason for this discrepancy? First, we know that the real atmosphere is not entirely opaque to longwave radiation and some part of it is transmitted to space. This suggests that the simple model overestimates the greenhouse effect of Earth's atmosphere. But the main reason for the overestimation of surface temperatures is the neglect of vertical heat transport by the atmospheric motions. However, this simple model captures the first order vertical structure of the atmosphere and shows how absorption of longwave radiation in the atmosphere (i.e. the greenhouse effect) leads to warmer surface temperatures and to a decrease of temperatures in the atmosphere with height.

The most accurate, but also the most computationally expensive, models of radiative transfer are lineby-line models, which calculate transmission, absorption, emission and scattering of radiation for each absorption line of many different gases in the atmosphere. Those models are complex and in climate models they are often replaced by simpler models that calculate radiative transfer in broader frequency bands. Those models confirm qualitatively the results from our simple 2 layer model that temperature decreases with altitude and that radiative transfer alone leads to surface temperatures warmer than those observed. In the real atmosphere such warm surface temperatures would trigger convective instability since the air in contact with the ground becomes lighter than the air aloft, which would lead to vertical motion. This process can be included in the model e.g. by limiting the maximum lapse rate to the observed global mean value of 6.5 K/km and vertically redistributing the heat required to do so. This is a radiative-convective model.

Radiative-convective models are useful to understand the effect of individual greenhouse gases on the vertical temperature structure. However, if the lapse rate is limited and a vertical energy transport due to convection is included both surface and upper tropospheric temperatures are in much better agreement with the observations. These calculations consider water vapor, CO_2 and ozone as radiatively active gases. The CO_2 mixing ratio is assumed constant whereas for water vapor and ozone fixed vertical distributions based on observations are used.

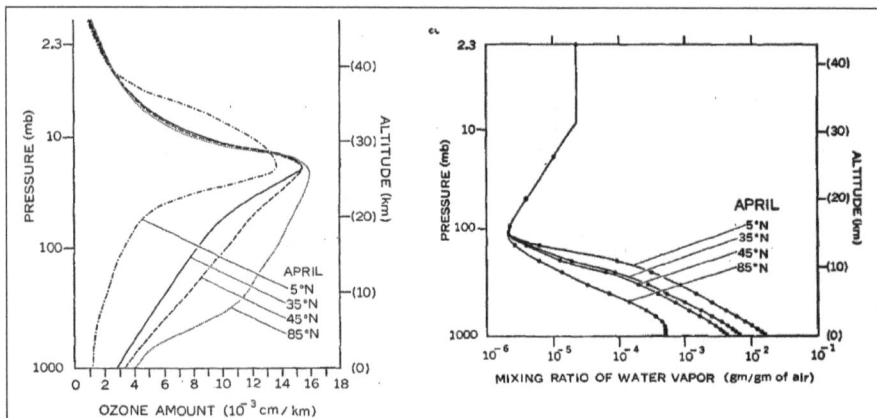

Vertical distribution of ozone (left) and water vapor (right) used in Manabe and Strickler (1964).

The left panel of figure shows that a CO_2 level of 290 ppm leads to ~ 10 K warmer surface temperatures than an atmosphere without CO_2. Absorption of solar radiation by ozone leads to constant temperatures in the tropopause and increasing temperatures in the stratosphere. The net heat loss due to radiative fluxes in the troposphere is balanced by heat gain through convection (right panel Figure). Water vapor is the dominant greenhouse gas in the troposphere. However, because its mixing ratio decreases quickly with height it is less important in the stratosphere, where the heat balance is dominated between radiative cooling by longwave emission of CO_2 and heating by solar absorption of ozone.

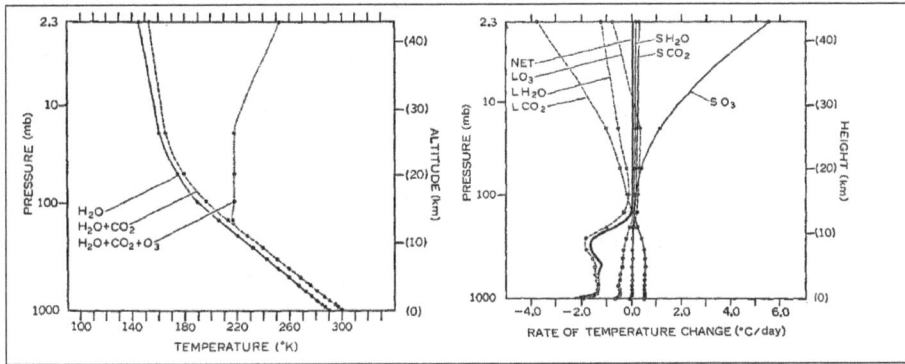

Left: The effect of CO_2 and ozone on the vertical temperature distribution in a radiativeconvective model.
Right: Heating rates associated with longwave (L) and shortwave (S) radiative fluxes.
The thick line is the net flux. From Manabe and Strickler (1964).

Because of this balance an increase in atmospheric CO2 leads to a cooling of the stratosphere. The current cooling trend in the stratosphere is probably caused by both, decreasing ozone and increasing CO_2.

Clouds affect longwave and shortwave fluxes. They reflect solar radiation very efficiently, which cools the surface during the daytime. But they are also almost perfect absorbers of longwave radiation, which warms the surface, particularly at night. Which of these two opposing effects wins depends on the height of the cloud, its albedo and its thickness.

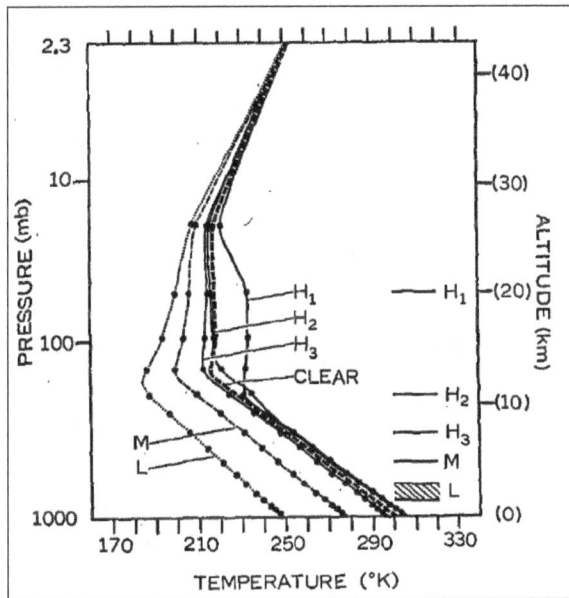

Effect of clouds on the vertical temperature distribution. From Manabe and Strickler (1964).

In radiative-convective models low clouds cool the surface and high clouds warm the surface. The lower the cloud top the smaller the longwave warming effect because the cloud will emit radiation at a temperature similar to the surface, whereas much of the solar radiation is reflected back to space. High clouds however, emit longwave radiation at much colder temperatures, which increases the greenhouse effect.

We can understand this a little more quantitatively by considering the change of the radiative balance at the top of the atmosphere if a cloud is added. Let's assume the cloud will have an albedo of a_{cloud} whereas the clear sky albedo is a_{clear}. The difference in shortwave radiation will thus be

$$\Delta F_{SW} = S\left(1-a_{cloud}\right) - S\left(1-a_{clear}\right) = -S\left(a_{cloud} - a_{clear}\right) = -S\,\Delta a \le 0$$

Let's also assume that the cloud top is above most of the longwave absorbing gas (water vapor), which limits the validity of our model to above ~ 4 km. In this case the longwave emission to space can be approximated by blackbody radiation from the cloud top $F_{LW} = \sigma T_{ct}^4$, so that the difference in longwave radiation at the top of the atmosphere becomes

$$\Delta F_{LW} = \sigma T_{ct}^4 - F_{LWclear}.$$

Thus the change in longwave flux will be negative if the cloud top is higher than the effective height of the clear sky longwave emission. The total change in the radiative balance at the top of the atmosphere becomes

$$\Delta R_{TOA} = \Delta F_{SW} - \Delta F_{LW} = -S\Delta\alpha + F_{LWclear} - \sigma T_{ct}^4.$$

Assuming a constant lapse rate we can replace the temperature of the cloud top with its height

$$T_{ct} = T_s - \Gamma_{z_{ct}}.$$

Figure shows the cloud radiative forcing at the top of the atmosphere according to equation $\Delta R_{TOA} = \Delta F_{SW} - \Delta F_{LW} = -S\Delta\alpha + F_{LWclear} - \sigma T_{ct}^4.$ for a solar flux of 342 Wm⁻², a clear-sky outgoing longwave flux of 265 Wm⁻², a surface temperature of 288 K and a lapse rate of 6.5 K/km. Positive values will lead to a warming and negative values will lead to a cooling.

Table: Cloud Radiative Forcing as Estimated from Satellite Measurements. From Harrison et al.

	Average	Cloud-free	Cloud forcing
OLR	234	266	31
Absorbed solar radiation	239	288	− 48
Net radiation	5	22	− 17
Albedo	30%	15%	+ 15%

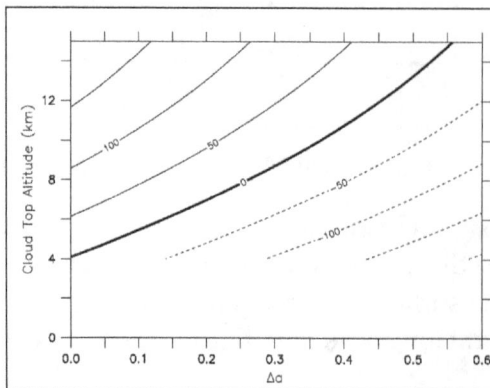

Cloud radiative forcing ΔR_{TOA} as a function of change in albedo and cloud top altitude. Negative values are show as dashed lines.

According to Figure high clouds with a low albedo lead to a warming, whereas low clouds with high albedo lead to a cooling.

A Simple Model of the Hadley Circulation

In 1735 George Hadley proposed that strong solar heating in the tropics causes air to rise. Close to the surface air must therefore flow towards the equator, whereas aloft the air must flow poleward. Hadley thought that this circulation cell extended all the way to the poles, but observations quickly showed that the extend of the cell is limited to the tropics.

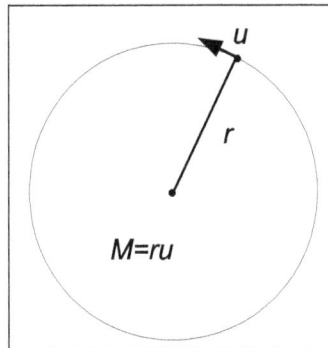

Held and Hou proposed a simple model to understand why the width of the Hadley cell is limited. Assume a 2-level atmosphere with an upper frictionless layer in which angular momentum M is conserved determining the zonal velocity $u = u_M$. Remember that the angular momentum per unit mass of an air parcel is its velocity times its radius r from the axis of rotation $M = ru$.

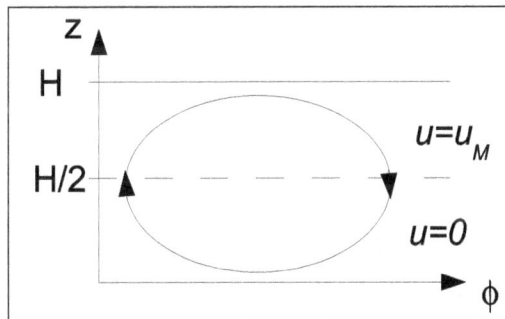

Two layer model of the Hadley cell. A frictionless upper layer is assumed in which the zonal velocity $u = u_M$ is determined by angular momentum conservation. In the lower layer friction leads to $u = 0$.

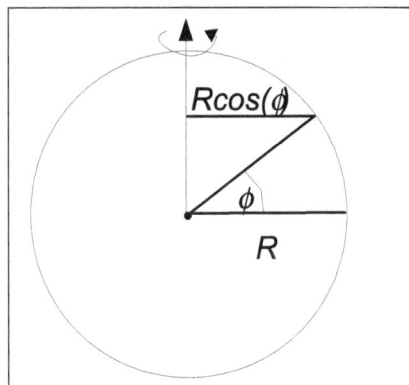

The lower layer is assumed to be dominated by friction such that the zonal velocity is zero. Heating near the equator will set up a clockwise circulation as depicted in figure with rising air near the equator and subsidence at higher latitudes. Newtonian relaxation of potential temperature towards some radiative equilibrium temperature Θ_E is assumed for the heating:

$$\frac{\partial \Theta}{\partial t} = \frac{\Theta - \Theta_E}{\tau_E},$$

With $\Theta_E = \Theta_0 - \Delta\Theta \left(3\sin^2(\phi) - 1\right)/3$. For small ϕ this leads to $\Theta_E = \Theta_{E0} - \Delta\Theta y^2 / R^2$, where y is the distance from the equator. Conservation of angular momentum (per unit mass) M is assumed as well as no zonal motion at the equator $u(\phi = 0) = 0$. At the equator the velocity of an air parcel is that of the solid Earth $(\Omega R = 462 \text{ m/s})$, its angular momentum is also that of the rotating Earth:

$$M = \Omega R^2,$$

Where $\Omega = 2\pi / 24h = 7.3 \cdot 10^{-5}$ m/s is the angular velocity of the Earth and $R = 6370$ km is Earth's radius. At latitude ϕ the zonal velocity will be that of the solid Earth plus that of the air relative to the Earth and its angular momentum will be

$$M = \left(\Omega R \cos\phi + u\right) R \cos\phi$$

From eqs. $M = \Omega R^2$, and $\Delta F_{SW} = S\left(1 - a_{cloud}\right) - S\left(1 - a_{clear}\right) = -S\left(a_{cloud} - a_{clear}\right) = -S\Delta a \leq 0$ it follows that the zonal wind increases with latitude as

$$u_M = \Omega R \frac{\sin^2 \phi}{\cos\phi} \simeq \frac{\Omega}{R} y^2.$$

Equation $u_M = \Omega R \frac{\sin^2 \phi}{\cos\phi} \simeq \frac{\Omega}{R} y^2$. predicts a strong increase of u_M with latitude such that at 30° the wind would blow at 110 m/s eastward. This is much stronger than the observed maxima in the jet stream. Turbulence and eddy activity leads to dissipation of potential vorticity, such that the assumption of no friction is no longer valid. Nevertheless, angular momentum conservation explains to a first order the acceleration of the zonal wind in the upper atmosphere - similar to a spinning ice dancer who draws her arms towards the body and spins faster.

The width of the Hadley cell can be determined by considering the vertical shear

$$\frac{\partial u}{\partial z} = \frac{u_M - 0}{H} = \frac{\Omega}{RH} y^2,$$

and the thermal wind balance

$$2\Omega \sin\phi \frac{\partial u}{\partial z} = \frac{g}{\Theta_0} \frac{\partial \Theta}{\partial y},$$

which leads to

$$\frac{\partial \Theta}{\partial y} = \frac{-2\Omega^2\Theta_0}{R^2 g H} y^3,$$

Or

$$\Theta_M = \Theta_{M0} - \frac{\Omega^2\Theta_0}{2R^2 g H} y^4.$$

The integration constant   can be determined through an energy (temperature) conservation argument. If we require no net heating the areas between the two curves in Fig. upper panel must be equal or

$$\int_0^{y_p} \Theta dy = \int_0^{y_p} \Theta_E dy.$$

Using also $\Theta_M\left(y_p\right)=\Theta_E\left(y_p\right)$ it follows that the width of the Hadley cell is

$$y_p = \left(\frac{\Delta\Theta g H 5}{\Omega^2\Theta_0 3}\right)^{1/2}$$

And

$$\Theta_{E0} - \Theta_{M0} = \frac{\Delta\Theta^2 g H 5}{R^2\Omega^2\Theta_0 18}$$

Using $\Delta\Theta = 100K$, $\Theta_0 = 288K$, and H = 8 km the width of the Hadley cell becomes $y_p = 2.9 \cdot 10^6$ m or about 3000 km or 30°, which is in good agreement with observations. However, the model by Held and Hou predicted a much too slow circulation.

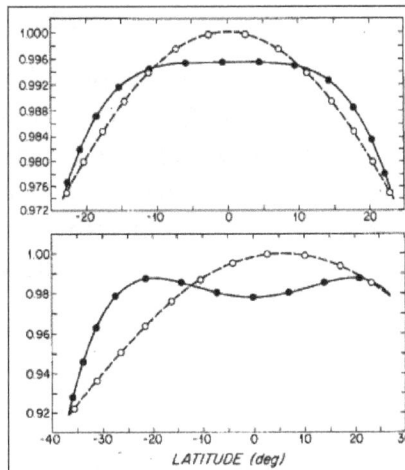

Radiative equilibrium temperature Θ_E / Θ_0 (dashed) and model temperature Θ / Θ_0 (solid) for the Held and Hou (1980) model with maximum heating at the equator ($\phi_0 = 0$, top panel) and the Lindzen and Hou (1988) model with maximum heating shifted slightly north of the equator ($\phi_0 = 6°$, bottom panel).

Lindzen and Hou (1988) showed subsequently that the reason for the underestimated circulation was the neglect of the seasonal cycle by Held and Hou. Including the seasonal cycle by shifting the latitude of maximum heating ϕ_0 only slightly away from the equator leads to a strong increase of the circulation strength. A shift by 4° increases the circulation by a factor of 100! The circulation is proportional to the heating $\Theta - \Theta_E$, which increases strongly for a slight shift in the latitude of maximum heating from the equator. This result again highlights the non-linear behavior of the climate system. Accounting for the seasonal cycle leads to an annually averaged circulation much stronger than one forced by annual averaged heating.

Figure shows the Hadley circulation and the zonal mean flow in the atmosphere from a reanalysis and the coarse resolution OSUVic climate model. The reanalysis is a global weather prediction model that has been run for a long time (40 years for the NCEP reanalysis) assimilating many observations that are routinely used for weather prediction. Although it is a model product it can be assumed to be an approximation of the real world. Some variables (e.g. temperature, pressure and velocities), however, are predicted better than others (e.g. precipitation). The reanalysis shows rising motion in the summer hemisphere near the equator and sinking motion between 15° and 30° in the winter hemisphere.

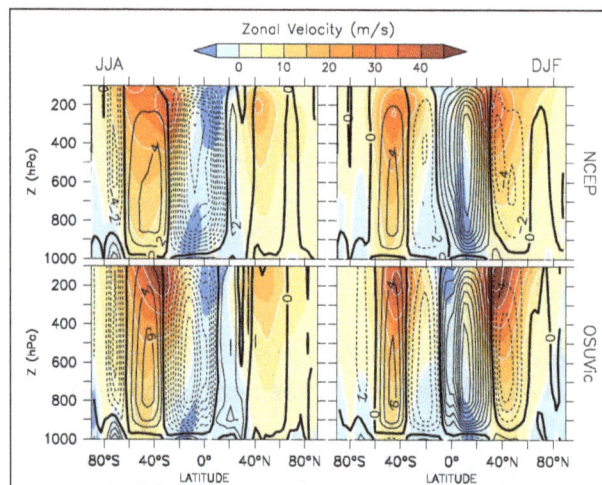

Figure: Atmospheric circulation for boreal summer (June-July-August, left) and boreal winter (December-January-February, right) from the NCEP reanalysis (top) and the OSUVic model (bottom). Colors show zonal wind velocities (ms^{-1}), with blue indicating easterlies and orange and red westerlies and 5 ms^{-1} isotach difference. White lines are drawn every 10 ms^{-1} isotach difference. Black contour lines show the meridional streamfunction (1010 kgs^{-1}), where positive/negative values (solid/dashed lines) indicate clockwise/counterclockwise flow.

The Hadley circulation has important implications for the hydrological cycle. Rising motions near the equator are associated with deep convection, release of latent heat and intense precipitation. Dry and cold air flows poleward in the upper troposphere. Subsidence in the subtropics leads to dry conditions near the surface. The is the reason why most deserts are located at subtropical latitudes and why the surface ocean salinity is high in the subtropics. The equatorward flow near the surface picks up water vapor from evaporation at the surface and moves it into the Intertropical Convergence Zone (ITCZ). Thus there is divergence of meridional water vapor transport in the subtropical atmosphere and convergence in the tropics.

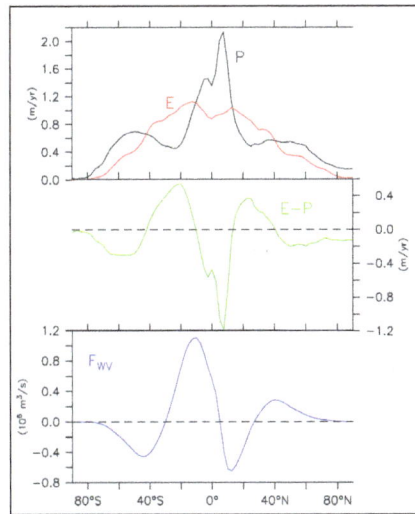

Zonally averaged evaporation (E) and precipitation (P).

General Circulation Models

General Circulation Models of the atmosphere solve conservation equations for momentum, mass, energy, water and possibly others (e.g. aerosols). The most commonly used set of equations is introduced below.

The Primitive Equations

The momentum equations are based on Newton's second law

$$\frac{d\vec{u}}{dt} = \sum \vec{F},$$

Which states that the change in momentum (velocity times mass) of an object with time is due to the sum of the forces acting on it. Equation $\frac{d\vec{u}}{dt} = \sum \vec{F}$, results from dividing by the mass, so that the forces on the right hand side are in forces per unit mass. The vectors are three dimensional. The time derivative in equation $\frac{d\vec{u}}{dt} = \sum \vec{F}$, is the total (Lagrangian) derivative moving with the object. The Navier-Stokes equations govern the motions of a fluid. In the case of an incompressible Newtonian fluid the NavierStokes equations are

$$\frac{d\vec{u}}{dt} = \underbrace{\frac{\partial \vec{u}}{\partial t} + \vec{u}\vec{\nabla}\vec{u}}_{\text{inertia}} = \underbrace{\frac{-1}{\rho}\vec{\nabla}p}_{\text{pressure gradient}} + \underbrace{\mu\nabla^2\vec{u}}_{\text{viscosity}} + \underbrace{f}_{\text{other body forces}} .$$

Now the total derivative

$$\frac{d}{dt} = \frac{\partial}{\partial t} + \vec{u}\vec{\nabla}$$

is split into the (local) change at a fixed location (partial derivative) and the advection (of momentum) with the fluid (second term). Both terms are called the inertia. The terms on the right hand side are the pressure gradient force, the frictional force due to viscosity (μ) and other body forces. On a rotating sphere Coriolis and centripetal forces enter the momentum equations, which, in spherical coordinates become:

$$\frac{du}{dt} - \left(f + u\frac{\tan\phi}{R} \right)v = -\frac{1}{\rho\, R\cos\phi}\frac{\partial p}{\partial\lambda} + F_u$$

$$\frac{dv}{dt} + \left(f + u\frac{\tan\phi}{R} \right)u = -\frac{1}{\rho R}\frac{\partial p}{\partial\phi} + F_v$$

$$g - \frac{1}{\rho}\frac{\partial p}{\partial z},$$

where the zonal, meridional and vertical components of velocity are $\vec{u} = (u,v,w)$, $f = 2\Omega\sin\phi$ is the Coriolis parameter, (λ,ϕ,z) are longitude, latitude and depth (height), p is the pressure, ρ the density and R Earth's radius. In the vertical momentum equation $g - \frac{1}{\rho}\frac{\partial p}{\partial z}$, the inertial terms have been neglected because they are much smaller than the acceleration due to gravity g=9.81 ms-2 and the pressure gradient term.

Assuming an incompressible fluid the equation for the conservation of mass becomes

$$\frac{\partial\rho}{\partial t} = -\vec{\nabla}\left(\rho\vec{u}\right) = 0 \rightarrow \vec{\nabla}\vec{u} = 0.$$

The equation for the conservation of energy can be derived from the first law of thermodynamics to become an evolution equation for temperature:

$$c_v\frac{dT}{dt} = -p\frac{d}{dt}\left(\frac{1}{\rho}\right) + F_T,$$

The first term on the right hand side of equation $c_v\frac{dT}{dt} = -p\frac{d}{dt}\left(\frac{1}{\rho}\right) + F_T$ is due to adiabatic expansion/compression of the fluid and the second term represents all diabatic processes, such as radiation or latent heat release during condensation of water vapor. In climate models equation

$c\frac{dT}{dt} = -p\frac{d}{dt}\left(-\right) + F$ is often replaced by an equation for the potential temperature

$$\Theta = T\left(\frac{p_0}{p}\right)^{\kappa} \rightarrow \frac{d\Theta}{dt} = F_\Theta,$$

which results in the cancelation of the adiabatic expansion/compression term $\kappa = R'/c_p$ with R' the gas constant of dry air and c_p the heat capacity at constant pressure. The potential temperature is the temperature an air or water parcel would have if adiabatically brought to the surface.

The equation for the conservation of water vapor can be written as an evolution equation for specific humidity q, which is the mass of water vapor per mass of moist air

$$\frac{dq}{dt} = F_q,$$

where F_q includes precipitation and evaporation. Precipitation is usually calculated as the excess water vapor above a certain threshold for relative humidity (typically about 80%). Relative humidity is defined as the specific humidity divided by its saturation concentration. The saturation specific humidity depends exponentially on temperature (Clausius-Clapeyron relation).

The equation of state relates density, temperature and pressure. For air it is the ideal gas law

$$p = \rho R'T$$

and for sea water it is a non-linear empirical function including salinity

$$\rho = \rho(p,T,S)$$

Equations $\frac{du}{dt} - \left(f + u\frac{\tan\phi}{R}\right)v = -\frac{1}{\rho R\cos\phi}\frac{\partial p}{\partial \lambda} + F_u$, $p = \rho R'T$, $\rho = \rho(p,T,S)$ comprise a set of seven equations with seven unknowns (u, v, w, ρ, p, T, q), known as the primitive equations, which can be solved principally if assumptions on F_u, F_v, F_T and F_q are made and if boundary conditions are specified.

Surface Processes

Surface fluxes in GCMs are calculated according to empirical relations, so called bulk formulae. Momentum fluxes (wind stress) are:

$$F_u = \rho C_m |\vec{u}| u,$$

$$F_v = \rho C_m |\vec{u}| v,$$

the sensible heat flux

$$F_\Theta = \rho c_p C_\Theta |\vec{u}| (\Theta_s - \Theta_a),$$

and the moisture flux

$$F_q = \rho C_q |\vec{u}| (q_s - q_a).$$

The transfer coefficients for momentum C_m, which are also called drag coefficients, for heat C_T and moisture C_q are in the order of 10^{-3} and depend on surface roughness, the stability of the boundary layer and other surface properties such as soil moisture, vegetation or snow cover. Boundary layer theory (Monin-Obukhov similarity theory) can be used to calculate them. Θ_s and q_s are the surface values of temperature and specific humidity, whereas Θ_a and q_a are the values in the atmosphere, typically at 10 m. Over sea ice covered oceans the heat and moisture fluxes are multiplied by (1-a), where a is the fraction of the grid cell area covered by sea ice. The surface specific humidity is the saturation specific humidity $q_{sat}(T)$ over the ocean.

Moist Processes

The saturation specific humidity depends exponentially on temperature according to the Clausius-Clapeyron relation

$$q_{sat} = 0.622\frac{p_{ws}}{p_a - p_{ws}},$$

Where

$$p_{ws} = e^{77.345+0.0057T-7235/T} / T^{8.2}$$

is the water vapor partial pressure, pa is the air pressure and the temperature T is in Kelvin. Figure shows that q_{sat} approximately doubles for each 10 K increase in temperature. The Clausius-Clapeyron relation can be derived from classical thermodynamics.

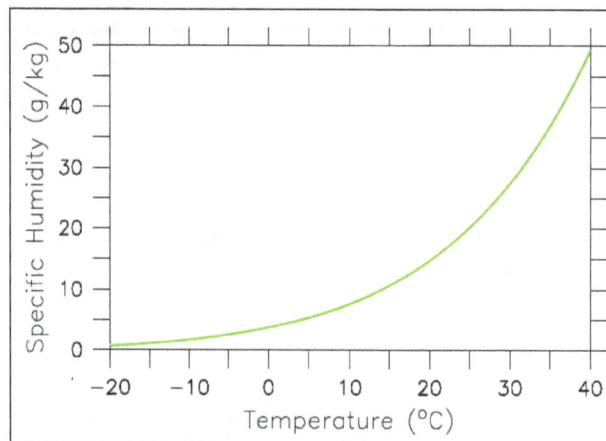

Saturation specific humidity for air as a function of temperature at 10^5 Pa atmospheric pressure.

The saturation specific humidity can be used to calculate precipitation. A simple way is to determine the excess specific humidity above a certain threshold (typically 80%) for relative humidity $rh = q/q_{sat}$. Precipitation is also associated with the release of latent heat of condensation or fusion depending on the ambient temperature. In the real world condensation happens only at 100% relative humidity but since the model grid scale is much larger than individual rain clouds it is warranted to use a smaller threshold because larger volumes of the atmosphere are usually a mix of clouds and cloud-free air and thus are rarely at 100% relative humidity.

Parameterizations

GCM grid boxes are typically several hundred kilometers in size and therefore equations $\frac{du}{dt} - \left(f + u\frac{\tan\phi}{R}\right)v = -\frac{1}{\rho R\cos\phi}\frac{\partial p}{\partial\lambda} + F_u$ - $p = \rho R'T$ $\rho = \rho(p,T,S)$ have to be averaged over large areas. Processes on spatial scales smaller than the grid scale must be described by formulas that include resolved quantities. Consider the advective flux uC of property C (e.g. temperature or specific humidity) with the velocity u. These terms occur e.g. on the left hand side of equations

$$c_v \frac{dT}{dt} = -p \frac{d}{dt}\left(\frac{1}{\rho}\right) + F_T \text{ and } \frac{dq}{dt} = F_q \text{ . Each variable is composed of its mean over the grid cell and}$$

the deviation from the mean

$$u = \bar{u} + u' \text{ and } C = \bar{C} + C', \text{ with } \overline{C'} = 0.$$

This is called Reynolds decomposition.

The mean flux becomes

$$\overline{uC} = \overline{\left(\bar{u} + u'\right)\left(\bar{C} + C'\right)} = \overline{\bar{u}\bar{C}} + \underset{0}{\underline{\overline{\bar{u}'C'}}} + \underset{0}{\underline{\overline{u'\bar{C}}}} + \overline{u'C'} = \bar{u}\bar{C} + \overline{u'C'},$$

where the first term on the right hand side is the flux due to the mean flow and the second term is the flux due to sub-grid scale fluctuations. These sub-grid scale fluxes are expressed using (turbulence) theory and/or empirical formulas.

In global coarse-resolution ocean models, for example, mesoscale eddies, which are in the order to tens to a hundred kilometers in size, are not resolved. Sub-grid scale fluxes are typically parameterized as a diffusive process

$$\overline{u'C'} = -K_C \frac{\partial \bar{C}}{\partial x},$$

which is proportional to the large scale gradient in C, where K_C is the diffusivity. (If C is velocity K_C = v is called viscosity).

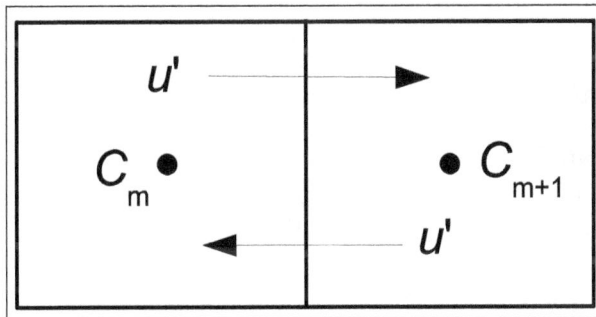

Consider two adjacent grid cells with concentrations C_m and C_{m+1} and, for simplicity no mean flow. If a sub-grid scale eddy transports fluid in the upper portion of the cell to the left and in the lower portion of the cell to the right with the same fluctuation velocity u', then the flux into cell m will be $\sim u'\Delta C$, where $\Delta C = C_{m+1} - C_m$. These sub-grid scale fluctuations will lead to down-gradient tracer flux consistent. We would expect the diffusivities $K_C \sim u'\Delta x$ to be proportional to the velocity fluctuations and the grid spacing.

Mixing along density surfaces (isopycnals) is more vigorous than across them, because it does not require energy input, whereas energy input is needed to mix across isopycnals. In modern ocean GCMs the diffusivity is therefore calculated separately for the along isopycnal directions and the across isopycnal (diapycnal) direction. Because density surfaces are mostly not equal to depth surfaces a diffusivity tensor is calculated. Along isopycnal diffusivities are typically on the order of 1000 m²/s whereas diapycnal diffusivities are around $10^{-5} - 10^{-4}$ m²/s. Diapycnal mixing may be caused by breaking of internal waves. Internal waves are waves in the ocean interior, in contrast to surface waves. Tidal flow over rough topography is one source for internal waves. Some waves break close to the generation sites, whereas other propagate away, which leads to mixing elsewhere. Mixing parameterizations considering the local generation and breaking of internal waves generated by tidal motions.

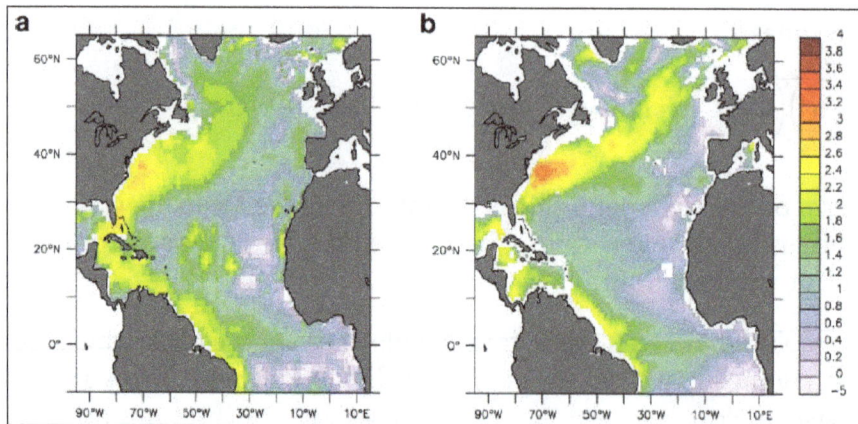

Logarithm of eddy kinetic energy calculated in a coarse resolution model (a) and from an eddy-resolving model (b) at 300 m depth in log10(EKE/[cm² s⁻²]).

Initially diffusivities in ocean models were constant. However, the velocity fluctuations are unlikely to be constant. Mesoscale eddy activity in the real ocean and in eddy-resolving models is not uniform but higher in certain regions, such as the Southern Ocean or the western boundary currents, and lower elsewhere. New and innovative approaches have been taken recently to account for the spatial distribution of mesoscale eddies. By solving for an additional equation for the eddy kinetic energy EKE it is possible for coarse-resolution models to reproduce the spatial variability of the mesoscale eddy field and its effect on sub-grid scale mixing.

Atmospheric models resolve the large scale baroclinic eddies that are associated with the high and low pressure systems and the daily fluctuations of weather. That is because the Rossby radius of deformation is much larger for the atmosphere than for the ocean. Thus fluxes associated with these transient eddies are resolved in global atmospheric GCMs, whereas they have to be parameterized in ocean GCMs. In atmospheric GCMs processes such as convection, clouds and precipitation are important at sub-grid scales and need to be parameterized.

Non-linear Dynamics and Chaos

The primitive equations are a set of non-linear coupled differential equations. Non-linear terms are e.g. in the advection terms, where products of velocities with gradiens of velocities or gradients of temperature or moisture occur. One consequence of these non-linear terms is that perturbations can grow and small differences in initial conditions can lead to large differences in the state of the system after some time.

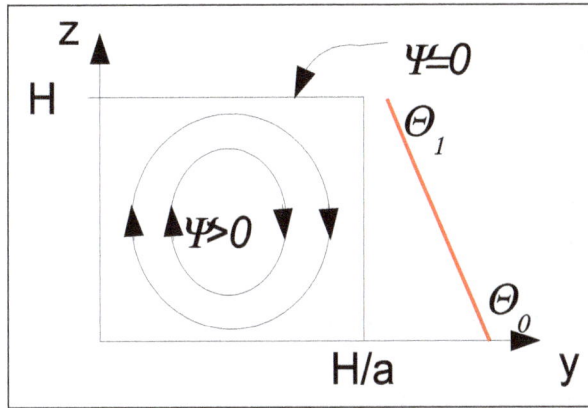

Schematic of the Lorenz-Saltzman model. A circulation in the 2-dimensional y-z
plane forced by a constant vertical temperature gradient (red line) is assumed.

Lorenz studied an approximation of the basic equations of convection in a 2-dimensional incompressible, viscous fluid in a non-rotating reference frame derived by Barry Saltzman. A
vertical overturning cell and a background constant vertical temperature gradient is assumed
as depicted in Figure. The basic equations are the incompressible version of the mass conservation equation

$$\frac{\partial v}{\partial y} + \frac{\partial w}{\partial z} = 0,$$

the equations for horizontal and vertical momentum

$$\frac{Dv}{Dt} = -\frac{1}{\rho_0}\frac{\partial p}{\partial y} + \mu\nabla^2 v$$

$$\frac{Dw}{Dt} = -\frac{1}{\rho_0}\frac{\partial p}{\partial z} + \mu\nabla^2 w - \frac{g}{\rho_0}\tilde{\rho},$$

where μ is a viscosity, $\tilde{\rho}$ is a small deviation from the constant density ρ_0, and

$$\frac{D}{Dt} = -\frac{\partial}{\partial t} + v\frac{\partial}{\partial y} + w\frac{\partial}{\partial z}$$

denotes the (Lagrangian) rate of change moving with the fluid.

Introducing the streamfunction

$$v = -\frac{\partial \psi}{\partial z}, \qquad w\frac{\partial \psi}{\partial y}$$

and the vorticity

$$\zeta = \frac{\partial w}{\partial y} - \frac{\partial v}{\partial z} = \nabla^2 \psi$$

the equation for the vorticity is derived by cross differentiating eqs. $\dfrac{Dv}{Dt} = -\dfrac{1}{\rho_0}\dfrac{\partial p}{\partial y} + \mu\nabla^2 v$ and $\dfrac{Dw}{Dt} = -\dfrac{1}{\rho_0}\dfrac{\partial p}{\partial z} + \mu\nabla^2 w - \dfrac{g}{\rho_0}\tilde{\rho}$.

$$\frac{D\zeta}{Dt} = \mu\nabla^2\zeta - \frac{g}{\rho_0}\frac{\partial\tilde{\rho}}{\partial y}.$$

With the definition of the expansion coefficient $\alpha = -\dfrac{1}{\rho_0}\dfrac{\partial\tilde{\rho}}{\partial\Theta}$ the vorticity equation

$$\frac{D\zeta}{Dt} = \mu\nabla^2\zeta + g\alpha\frac{\partial\Theta}{\partial y}$$

reveals that vorticity is produced by buoyancy forcing and destroyed by viscosity. Now we assume that the temperature distribution

$$\Theta = \Theta_0 - \frac{\Delta T}{H}z + \tilde{\Theta}(y,z,t)$$

is a small deviation $\tilde{\Theta}$ from a constant vertical gradient. Energy conservation allows for some diffusion

$$\frac{\quad}{Dt} = \nabla\,\Theta$$

Inserting equation $\Theta = \Theta_0 - \dfrac{\Delta T}{H}z + \tilde{\Theta}(y,z,t)$ in $\dfrac{D\Theta}{Dt} = \kappa\nabla^2\Theta$. and using the definition of the streamfunction $v = -\dfrac{\partial\psi}{\partial z},\ w\dfrac{\partial\psi}{\partial y}$ yields

$$\frac{\partial\tilde{\Theta}}{\partial t} - \frac{\partial\psi}{\partial z}\frac{\partial\tilde{\Theta}}{\partial\ddot{y}} + \frac{\partial\psi}{\partial y}\frac{\partial\tilde{\Theta}}{\partial z} = \kappa\nabla^2\tilde{\Theta} + \frac{\Delta T}{H}\frac{\partial\psi}{\partial y}.$$

From the vorticity equation $\dfrac{D\zeta}{Dt} = \mu\nabla^2\zeta + g\alpha\dfrac{\partial\Theta}{\partial y}$ together with $v = -\dfrac{\partial\psi}{\partial z},\ w\dfrac{\partial\psi}{\partial y}$ and $\zeta = \dfrac{\partial w}{\partial y} - \dfrac{\partial v}{\partial z} = \nabla^2\psi$ we get

$$\frac{\partial}{\partial t}\nabla^2\psi - \frac{\partial\psi}{\partial z}\frac{\partial}{\partial y}\nabla^2\psi\frac{\partial\psi}{\partial y}\frac{\partial}{\partial z}\nabla^2\psi = \mu\nabla^4\psi + g\alpha\frac{\partial\Theta}{\partial y}.$$

Equations $\dfrac{\partial\tilde{\Theta}}{\partial t} - \dfrac{\partial\psi}{\partial z}\dfrac{\partial\tilde{\Theta}}{\partial\ddot{y}} + \dfrac{\partial\psi}{\partial y}\dfrac{\partial\tilde{\Theta}}{\partial z} = \kappa\nabla^2\tilde{\Theta} + \dfrac{\Delta T}{H}\dfrac{\partial\psi}{\partial y}.$ and $\dfrac{\partial}{\partial t}\nabla^2\psi - \dfrac{\partial\psi}{\partial z}\dfrac{\partial}{\partial y}\nabla^2\psi\dfrac{\partial\psi}{\partial y}\dfrac{\partial}{\partial z}\nabla^2\psi$

$= \mu\nabla^4\psi + g\alpha\dfrac{\partial\Theta}{\partial y}$. are a coupled, non-linear system of partial differential equations. With appropriate boundary conditions those can be solved. The following Fourier expansion of the streamfunction and temperature deviation satisfies the boundary conditions of zero streamfunction and no horizontal gradient at y = 0 and y = H/a and $\tilde{\Theta}=0$ at z = 0 and z = H:

$$\psi(y,z,t)=X(t)\sin\left(\frac{\pi a y}{H}\right)\sin\left(\frac{\pi z}{H}\right)+....$$

$$\tilde{\Theta}(y,z,t)=Y(t)\cos\left(\frac{\pi a y}{H}\right)\sin\left(\frac{\pi z}{H}\right)-Z(t)\sin\left(\frac{2\pi z}{H}\right)+...$$

This choice allows solutions with the simplest possible spatial structure. But eqs.

$$\psi(y,z,t)=X(t)\sin\left(\frac{\pi a y}{H}\right)\sin\left(\frac{\pi z}{H}\right)+....$$ and

$$\tilde{\Theta}(y,z,t)=Y(t)\cos\left(\frac{\pi a y}{H}\right)\sin\left(\frac{\pi z}{H}\right)-Z(t)\sin\left(\frac{2\pi z}{H}\right)+...$$ are approximations because they

neglect higher order terms in the Fourier expansion. Inserting equations

$$\psi(y,z,t)=X(t)\sin\left(\frac{\pi a y}{H}\right)\sin\left(\frac{\pi z}{H}\right)+....$$ and

$$\tilde{\Theta}(y,z,t)=Y(t)\cos\left(\frac{\pi a y}{H}\right)\sin\left(\frac{\pi z}{H}\right)-Z(t)\sin\left(\frac{2\pi z}{H}\right)+...$$ in equations

$$\frac{\partial\tilde{\Theta}}{\partial t}-\frac{\partial\psi}{\partial z}\frac{\partial\tilde{\Theta}}{\partial\ddot{y}}+\frac{\partial\psi}{\partial y}\frac{\partial\tilde{\Theta}}{\partial z}=\kappa\nabla^2\tilde{\Theta}+\frac{\Delta T}{H}\frac{\partial\psi}{\partial y}.$$ and

$$\frac{\partial}{\partial t}\nabla^2\psi-\frac{\partial\psi}{\partial z}\frac{\partial}{\partial y}\nabla^2\psi\frac{\partial\psi}{\partial y}\frac{\partial}{\partial z}\nabla^2\psi=\mu\nabla^4\psi+g\alpha\frac{\partial\Theta}{\partial y}.,$$ and introducing a dimensionless time

$\tau=(\pi/H)^2(1+a^2)\kappa t$ it can be shown that the following set of ordinary differential equations emerges

$$\frac{dX}{dt}=-\sigma X+\sigma Y$$

$$\frac{dY}{dt}=-XZ+rX-Y$$

$$\frac{dZ}{dt}=XY-bZ,$$

Where $\sigma=\mu/\kappa$ is the Prandtl number, $r=R_c^{-1}R_a=g\alpha H^3\Delta T a^2\mu^{-1}\kappa^{-1}\pi^{-1}(1+a^2)^{-3}$, and $b=4(1+a^2)^{-1}$. In these equations X is proportional to the convective motion, while Y is proportional to the temperature difference between ascending and descending currents, and Z is proportional to the distortion of the vertical temperature profile from linearity.

Ocean

A Simple Box Model of the Thermohaline Circulation

Henry noted differences in air-sea interactions between temperature and salinity and explored the effect on ocean circulation in a simple 2 box model. The density of sea water is determined by both temperature and salinity such that cooler temperatures and higher salinities lead to heavier densities. Whereas surface salinity is influenced by evaporation (E) and precipitation (P), it does not, in itself, influence the surface fresh water fluxes (E-P). The temperature of the ocean surface is modulated by atmospheric heat fluxes, but it does also influences those heat fluxes, because both sensible and latent heat fluxes depend on the temperature difference between the ocean and the atmosphere. Thus there is a fundamental difference between the coupling of sea surface temperatures to the atmosphere (which is two way) and the coupling of sea surface salinity (which is one way). Stommel use a simple idealistic thought experiment to illustrate the is effect. He the assumed two vessels, each well mixed as indicated by the stirrers in figure.

The temperature T and salinity S of each vessel is controlled through a porous wall which is connected to another (infinite) container with constant temperature and salinity. Stommel was interested in the case in which the transfer of temperature through the wall is faster than the transfer of salinity. This represents strong coupling between surface ocean and atmospheric temperatures and weak influence of salinity on the fresh water flux. Here we assume that vessel 1 is kept at a lower temperature than vessel 2 with $\Delta T = T_1 - T_2 < 0$ and that the salt flux through the wall can be replaced by a constant surface freshwater flux F out of vessel 2 and into vessel 1 as indicated by the blue arrows in Figure.

The vessels are connected through a pipe (capillary) at the bottom and an overflow at the top. The overflow ensures equal fluid levels in both vessels. The flow through the capillary is determined by the density difference between the vessels. If the water in vessel 1 becomes denser than that in vessel 2 the hydrostatic pressure at the bottom of vessel 1 will be greater than that in vessel 2 and a flow through the capillary from vessel 1 to vessel 2 would result. At steady state this flow will be compensated by a return of water across the overflow.

The change in salinity with time in the two vessels is

$$\frac{\partial S_1}{\partial t} = |q|(S_2 - S_1) - F$$

$$\frac{\partial S_2}{\partial t} = |q|(S_1 - S_2) + F$$

The flow will be proportional to the density difference

$$q = c\Delta\rho = c(\rho_1 - \rho_2) = c(\alpha\Delta T + \beta\Delta S)$$

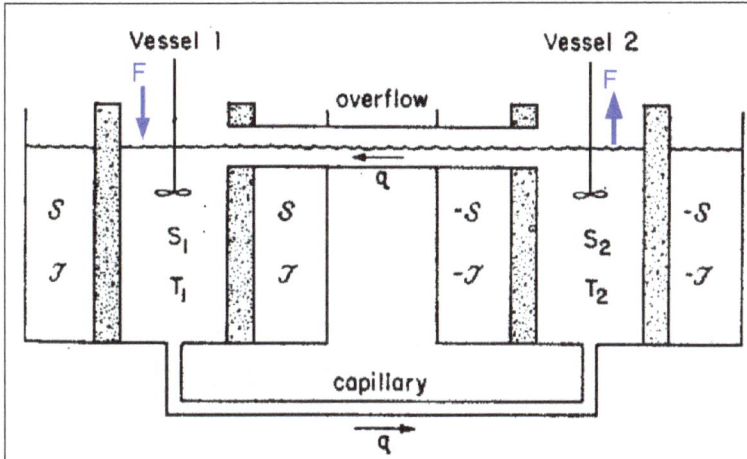

Thermohaline circulation between two vessels according to Stommel. The original model by Stommel has porous walls with transfer of heat and salt from an infinite outside reservoir. Here we replace the salinity transfer through the wall by a fixed surface freshwater flux F (blue arrows).

With $c > 0$ a constant depending on the viscosity through the pipe, $\alpha > 0$ the thermal expansion coefficient for sea water and $\beta > 0$ the haline contraction coefficient. Subtracting $\frac{\partial S_2}{\partial t} = |q|(S_1 - S_2) + F$ from $\frac{\partial S_1}{\partial t} = |q|(S_2 - S_1) - F$ give

$$\frac{\partial \Delta S}{\partial t} = 2|q|\Delta S + 2F$$

Thus at steady state the salinity difference $\Delta S_0 = -F / |q|$ is determined by the surface freshwater flux such that the vessel experiencing evaporation will be saltier than the one experiencing precipitation. But the difference will be reduced by the exchange flow between the vessels.

For q > 0 steady states are at

$$\gamma \equiv -\frac{\Delta S_0}{\tau} = \frac{1}{2} \mp \sqrt{\frac{1}{4} - \frac{\vec{F}}{\tau^2}},$$

Where $\tau = \frac{\alpha}{\beta}\Delta T > 0$ and $\tilde{F} = \frac{F}{c\beta}$. A stability analysis shows that the equilibrium is stable if $\gamma > 1/2$. This stable equilibrium is shown as the upper solid line in Figure.

For q < 0 steady states are easily found by replacing F by -F.

$$\gamma = \frac{1}{2} \mp \sqrt{\frac{1}{4} + \frac{\tilde{F}}{\tau^2}}$$

The stability analysis shows that states are stable if $\gamma < 1/2$. Thus only the minus sign in

$$\gamma = \frac{1}{2} \mp \sqrt{\frac{1}{4} + \frac{\tilde{F}}{\tau^2}} \text{ is}$$

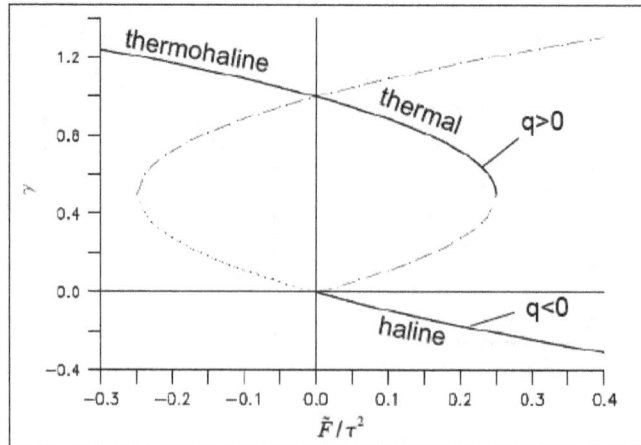

Equilibrium solutions for the Stommel model. The stable branches of the solution are shown
as the black heavy lines. The dashed and dotted lines show unstable solutions.

permitted and we also require $\tilde{F} > 0$. This part of the solution is shown as the lower solid line in
Figure. We realize that the system has two steady states for a certain range of the freshwater flux
$0 \le F \le c\alpha^2 \Delta T^2 / (4\beta)$. At both thresholds of the freshwater flux abrupt changes from one branch
of the stable solution to the other are possible.

Ocean General Circulation Models

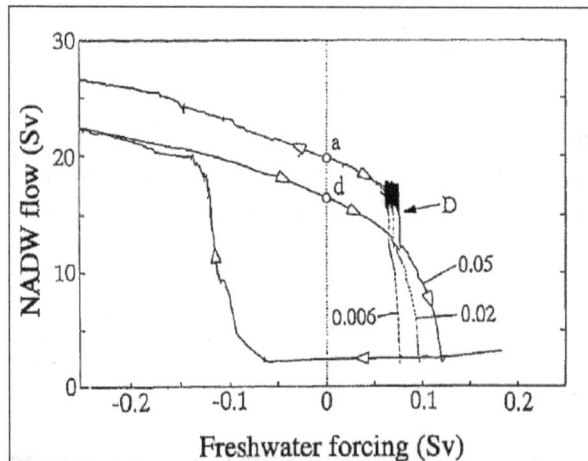

Hysteresis behavior of the Atlantic meridional overturning circulation (NADW) in an ocean
general circulation model forced with freshwater perturbation in the North Atlantic.

The first ocean GCMs have been developed in the 1960s (e.g. Bryan and Cox, 1967). Initially ocean
only models were forced by wind stress and restoring (nudging) temperature and salinity at the
surface to observed values. Restoring boundary conditions (BCs) for salinity were later replaced
by fixed salt (or freshwater) fluxes that could e.g. be determined from an initial simulation with
restoring BCs. Models with these so-called "mixed BCs" exhibited different states for the same BCs
and parameters, similar to Stommel's box model.

Today's models have a typical resolution of a few hundred kilometers and 20-40 vertical levels.
They simulate both the shallow wind-driven as well as the deep global overturning circulation
(Fig.). Most of the depth integrated flow, shown in Fig., is due to the circulation in the upper

ocean featuring prominently the subtropical gyres in all ocean basins (red arrows in Fig.). These gyres are driven by westerly (from the west) winds at mid latitudes and the easterly trade winds at low latitudes. This leads to westward flow at low latitudes and eastward flow at mid latitudes. The Gulf Stream (northward flow in the Atlantic along the east coast of North America), Kuroshio (northward flow in the Pacific along the east coast of Asia), and Humbolt Current (northward flow along the west coast of South America) are part of the subtropical gyre. Ekman drift, which leads to flow perpendicular (toward the right/left in the northern/southern hemisphere) to the wind, causes convergence and downwelling in the centers of the subtropical gyres and upwelling along the equator and in the Southern Ocean (not all of these features are visible in the Figures). Subpolar gyres (light blue arrows in Fig. 6.4) are simulated in the North Atlantic (counter clockwise) and Southern Ocean (Weddell and Ross Seas). The strongest current in the world ocean $(\sim 100\ \text{Sv})$ is the Antarctic Circumpolar Current (ACC) flowing eastward around Antarctica (blue arrows in figure).

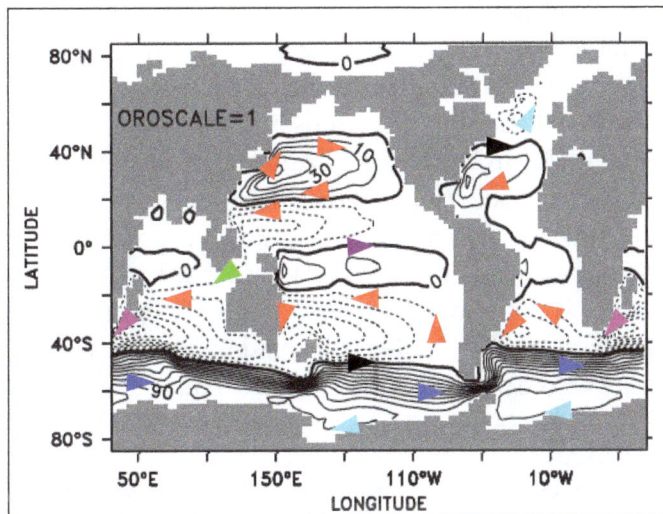

Barotropic (depth integrated) ocean circulation simulated by the OSUVic model. Contour lines of the streamfunction in Sv show clockwise (counter clockwise) flow for positive (negative=dashed) values.

Sinking of surface waters to the deep ocean at high latitudes and upwelling at low latitudes and in the Southern Ocean are the main features of the deep meridional overturning circulation (Fig.). Surface waters in the North Atlantic are dense because they are salty and cold, which causes them to sink and flow south at mid-depths (2-3 km) along the west coast of the Americas (purple arrows in Fig.). Eventually this North Atlantic Deep Water (NADW) flows into the Southern ocean, where it mixes with other water masses, some upwells to the surface and some flows into the Indian and Pacific oceans as Circumpolar Deep Water (CDW, green arrow in Fig.). There some upwells to the surface and returns to the North Atlantic via the Indonesian Throughflow (green arrow in Fig.) and Benguela Current around the tip of South Africa (purple arrows in Fig.).

All of these simulated features are realistic. However, due to the coarse resolution of the model narrow features (such as western and eastern boundary currents) are not well represented. Mesoscale eddies are not resolved in this nor in most other global ocean models and their effects are parameterized. Tides and gravity waves are also typically not simulated in GCMs. Some models also consider geothermal heat flux from the sea floor, which as been shown to intensify the meridional overturning circulation.

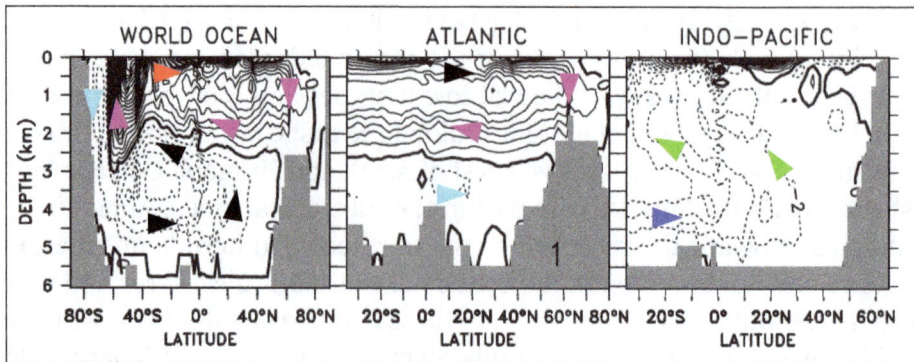

Meridional overturning circulation simulated by the OSUVic climate model. Streamfunction shown with isoline difference of 2 Sv. Solid (dashed) lines indicate clockwise (counter clockwise) flow. The zero line is bold.

We have shown recently that the observed pattern of the meridional overturning circulation with sinking in the North Atlantic but not in the North Pacific is due to the effect of mountains and ice sheets on land on atmospheric water vapor transport. The Atlantic is saltier than the Pacific even though more rivers enter the Atlantic than the Pacific. But water vapor is transported in the atmosphere from the Atlantic to the Pacific via trade winds blowing over the Panama isthmus. The ranges of the Rocky Mountains in North America and Andes in South America limit the transport of water vapor via the westerlies from the Pacific to the Atlantic. Thus the configuration of mountains leads to more evaporation and less rain and river runoff in the Atlantic compared to the Pacific. This makes the Atlantic saltier and promotes deep water formation there, whereas the North Pacific is very fresh surface waters are too buoyant to sink to great depths. If mountains and ice sheets are removed in the model no more deep water is formed in the North Atlantic, whereas salinities in the North Pacific increase to the point that sinking takes place there.

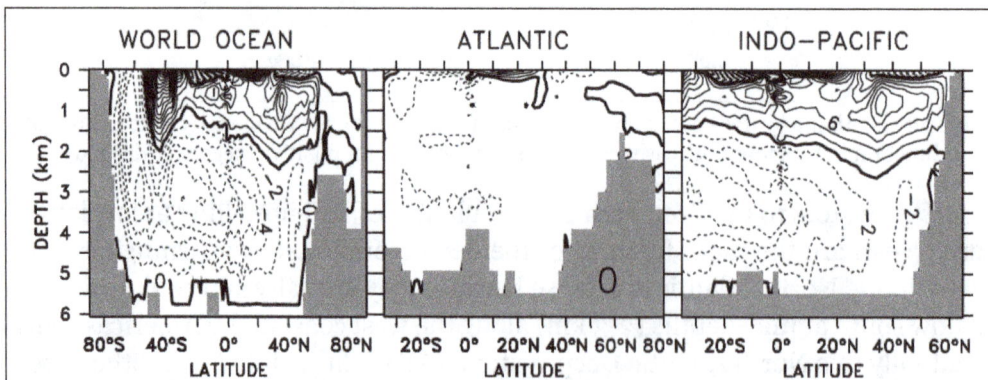

Meridional overturning circulation in a model without mountains and ice sheets.

Cryosphere

Sea Ice

Sea ice has two important effects. First, it insulates the ocean from the atmosphere. Heat transfer between the surface ocean and the atmosphere is strongly reduced in the presence of sea ice. Mass transfer is also blocked, such that air-sea gas and water exchange are strongly reduced. Second, sea ice formation, transport and melting is associated with a buoyancy flux to the surface ocean. Sea ice contains very little salinity (typically about 5 permil in contrast to sea water which has about 30 permil). This is because the freezing point of water decreases as its salinity increases – yes, that's the reason why we sprinkle salt on the road in winter. Thus during the freezing process only the freshwater freezes leaving the salt behind, which collects in pockets of brain water with very high salinity. Those pockets slowly melt their way down through the ice and eventually are released into the underlaying sea water. Thus, where sea ice freezes the ocean gets saltier and hence heavier. Wind and ocean currents move the ice around such that it typically melts at a different location than where it was formed. Melting leads to an input of freshwater and hence buoyancy into the surface ocean.

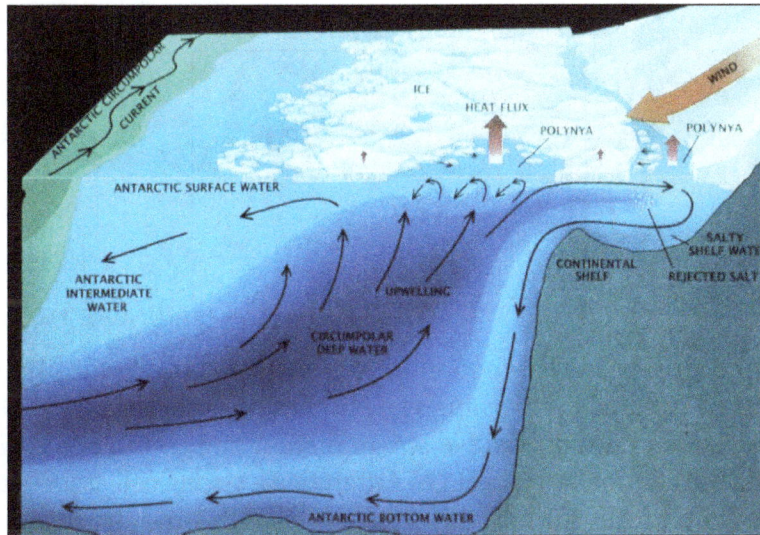

Figure: Sea ice effects on ocean and atmosphere in the Southern Ocean. Katabatic winds descending from the Antarctic ice sheet push sea ice offshore and create leads and polynyas where new sea ice can form. Formation of sea ice is associated with release of brine water with high salinities, thereby densifying the underlaying sea water. The cold and salty water sinks down on the continental shelf and flows across the shelf-break, along the continental slope into the abyssal ocean forming Antarctic Bottom Water.

Wind driven sea ice motion is very important for deep water formation in the Southern Ocean. Sea ice formation near Antarctica leads to salt input and increases Antarctic Bottom Water (AABW) formation. The westerly winds force northward Ekman transport of sea ice and melting in the areas of Antarctic Intermediate Water (AAIW) formation. AAIW is relatively fresh in part because of the associated input of freshwater from sea ice melting.

The penetration of anthropogenic chemicals such as chlorofluorocarbons (CFCs) can be used to monitor deep water formation. CFCs have been released into the atmosphere since the beginning of the 20th century due to their use in refrigerators and sprays. CFCs are now no longer used in most countries because they lead to ozone destruction in the stratosphere creating the harmful ozone hole. However, their penetration into the deep ocean reveals locations of deep water formation.

Figure shows elevated CFC concentrations observed near the sea floor of the Ross Sea suggesting AABW formation there. The model with wind driven sea ice reproduces these observations, whereas a model without sea ice motion produces no AABW in the Ross Sea and unrealistic downwelling in the Drake Passage.

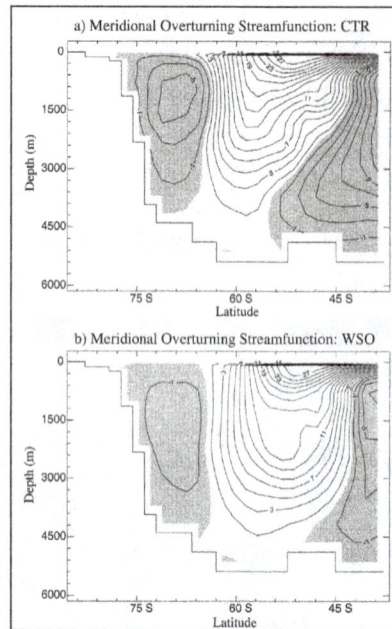

Figure: Effect of wind driven sea ice motion on the meridional overturning circulation of the Southern Ocean. The meridional overturning streamfunction is shown in units of Sv ($1Sv = 10^6 m^3/s$). Flow is clockwise along lines of positive streamfuction and counterclockwise along negative (shaded) lines. Along the continental margin of Antarctica a strong cell associated with about 5 Sv of Antarctic Bottom Water formation exists in the model with wind driven sea ice motion (top), whereas this cell is much weaker in the model without wind driven sea ice (bottom).

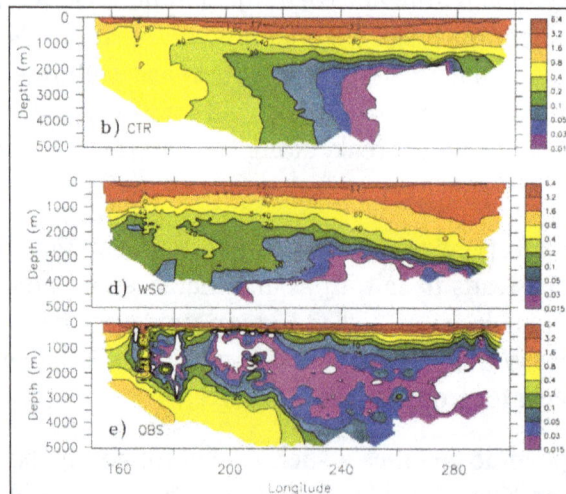

CFC concentration along an zonal section in the Pacific section.

A Simple Ice Sheet Model

Perfectly Plastic Solution for an Ice Sheet on a Flat Base

Assumptions:

- Flow is quasi-two dimensional.

- Normal stress deviations are small.

- The surface slope (s < 0.1) is small.

Ice flows because of shear forces, caused by gravity, lead to plastic (irreversible) deformation of the ice. The balance of forces within the ice is one where the vertical gradient of the shear stress is equal to the horizontal pressure gradient caused by the slope of the ice sheet surface.

$$\frac{\partial \tau_{xz}}{\partial z} = \rho g s \implies \tau_{xz} = \rho g (H - z) s$$

where H is the surface elevation. The stress at the base of the ice sheet will therefore be

$$\implies \tau_b = \rho g H s = \rho g H \frac{\partial H}{\partial x} = const. = \tau_0.$$

If the basal stress is horizontally constant we can integrate equation $\implies \tau_b = \rho g H s = \rho g H \frac{\partial H}{\partial x} = const. = \tau_0.$ to give

$$\frac{1}{2} \frac{\partial H^2}{\partial x} = \frac{\tau_0}{\rho g} \implies H = \sqrt{\frac{2\tau_0}{\rho g}} x = \Lambda \sqrt{x}.$$

The ice sheet profile is parabolic.

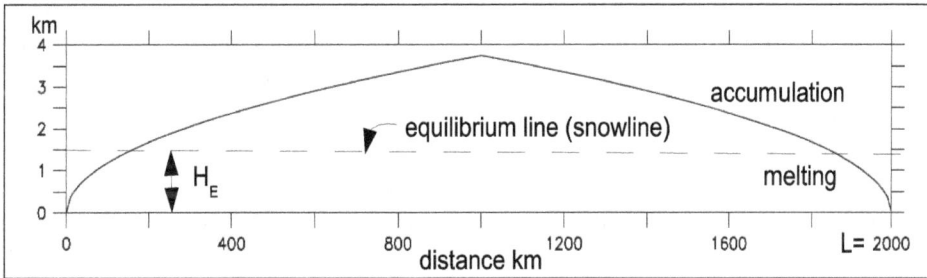

Ice sheet elevation H (equation 7.3) as a function of distance.

Because temperature decreases with height in the atmosphere higher parts of the ice sheet will have colder temperatures and therefore more likely a net positive surface mass balance. Let's assume that the surface mass balance increases (decreases) linear above (below) the equilibrium line H_E:

$$B = a(H - H_E).$$

The equilibrium line is to a large degree controlled by climate. In this case the total mass balance is positive when the mean ice sheet height is above the equilibrium line

$$\overline{H} = \frac{1}{L} \int_0^L H dx = \frac{2\Lambda}{3} \sqrt{L} > H_E$$

Thus an ice sheet with a mean hight above the equilibrium line tends to grow bigger, whereas an ice sheet with a mean height below H_E tends to melt away. Once the ice sheet has disappeared it will not grow again until the equilibrium line goes below zero. This positive feedback between ice sheet height and mass balance leads to hysteresis behavior and two equilibria. A warming climate, for example, can increase the equilibrium line latitude above the mean height of the ice sheet and induce its demise. Subsequent cooling and lowering of the equilibrium line does not bring the ice sheet back until H_E becomes negative.

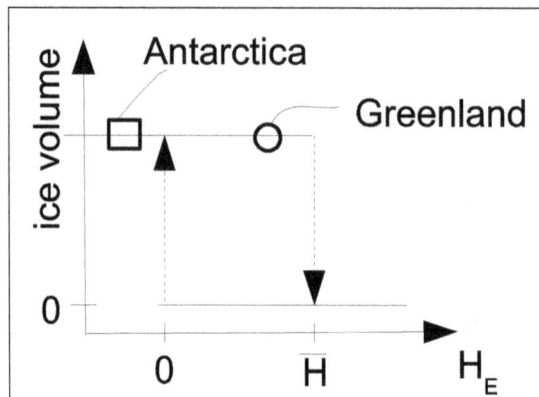

Hysteresis behavior of ice sheet volume as a function of equilibrium line altitude.

Of the two main ice sheets on Earth today, the one on Greenland is much closer to the threshold and therefore more sensitive to climate change, than the Antarctic ice sheet. The Antarctic ice sheet is very tall and located further poleward than the Greenland ice sheet. The equilibrium line for the Antarctic ice sheet is negative, that is, there is almost no mass loss due to surface melt. It looses most of its mass through calving of glaciers into the ocean. Therefore it has a much larger difference between the equilibrium line altitude and its mean height and is less vulnerable to climate change compared to the Greenland ice sheet, which is less tall and who's equilibrium line is closer to the mean height. Greenland looses a considerable amount of mass through surface melting. Indeed the observed increase of the surface melt area of the Greenland ice sheet in recent years is a matter of great concern with important implications for sea level rise. Greenland stores water equivalent of 7 m global sea level. Antarctica is good for about 70 m sea level rise.

Bedrock adjustment

Large ice sheets can be 3-4 km thick. The weight of such an ice sheet is enough to depress the lithosphere below. The continental crust floats on the upper mantle, which is partially melted and deformable (it flows very slowly). Thus the bedrock is not static but it sinks slowly in response to the weight of a large ice sheet. The bedrock adjustment can be included in ice sheet models by solving the following equation

$$\frac{\partial h}{\partial t} = \frac{(\rho_i / \rho_B) H - h}{\tau_B},$$

where h is the bedrock depression, $H = \tilde{H} + h$ is the ice sheet thickness and $\tilde{H}$ is the ice sheet elevation above the undisturbed bedrock. The time scale $\tau_B \sim 3-5$ ka is the response time of the bedrock.

A Numerical Model Using Glen's Law

$$\frac{\partial H}{\partial t} = \vec{\nabla}\overrightarrow{M} + G,$$

where G is the surface mass balance and

$$\overrightarrow{M} = AH^{m+1}\left|\vec{\nabla}\tilde{H}^{m-1}\right|\vec{\nabla}\tilde{H}$$

is the vertically integrated mass flow, which follows from Glen's law for the relation between vertical shear and stress $u = B\tau_B^m$ with m=3 and equation $\Rightarrow \tau_b = \rho g H s = \rho g H \frac{\partial H}{\partial x} = const. = \tau_0..$ With this we can rewrite equation $\frac{\partial H}{\partial t} = \vec{\nabla}\overrightarrow{M} + G,$ as

$$\frac{\partial H}{\partial t} = \vec{\nabla}\left(D\vec{\nabla}\tilde{H}\right) + G,$$

with the diffusivity

$$D = AH^{m+1}\left[\left(\frac{\partial\tilde{H}}{\partial x}\right)^2\left(\frac{\partial\tilde{H}}{\partial x}\right)^2\right]^{(m-1)/2}$$

Biosphere

Ocean Ecosystem and Carbon Cycle Models

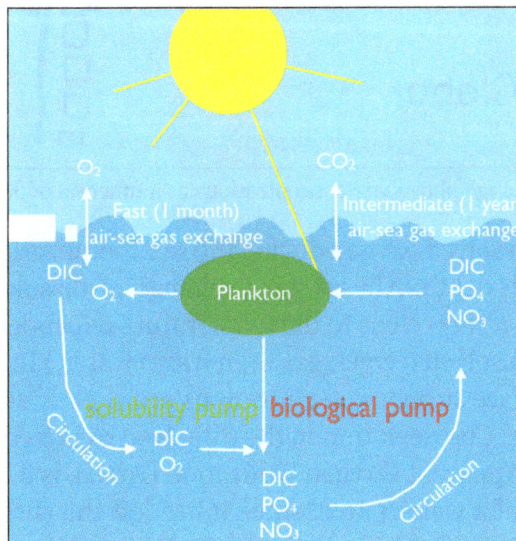

Schematic of ocean carbon, nutrient and oxygen cycles.

Biogeochemical processes in the ocean are governed by some of the same principles as on land. Photosynthesis converts inorganic carbon and nutrients into organic matter a process that produces oxygen, whereas whereas heterotrophic bacteria oxydize organic matter back to inorganic forms, a process that consumes oxygen. However, there are also important differences. Whereas on land leaves fall to the ground and nutrients and carbon is readily available for the next growth cycle, the sinking of organic matter to the deep sea makes nutrients at the surface scarce.

The light-filled upper ocean, the euphotic zone, where photosynthesis can take place is only about 120 m deep. The constant removal of nutrients from the euphotic zone by sinking organic matter would deplete the surface ocean quickly of all nutrients. However, upwelling and mixing of surface and subsurface waters brings back nutrients to the euphotic zone where they fuel plankton growth. Still, as a consequence of the sinking and remineralization of organic matter, global average surface nutrient concentrations are much smaller near the surface than at depths. In fact, in many regions of the surface oceans, such as the subtropical gyres, nutrient concentrations are below the detection limit for measurements. Surface nutrient concentrations are relatively large along the equator and eastern boundaries, where upwelling occurs and at high latitudes, where deep mixing occurs and where light is limited in winter.

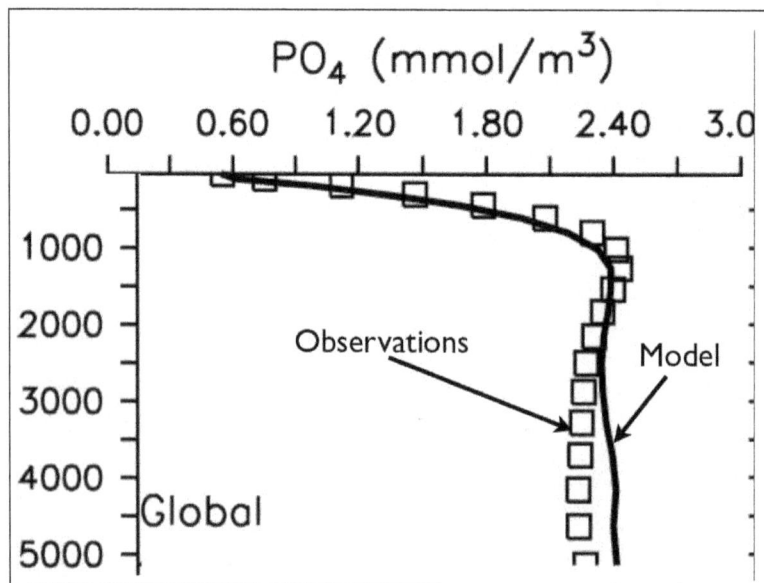

Global average phosphate concentrations as a function of depth (m).

Due to the consumption of dissolved oxygen during the remineralization of organic matter, oxygen concentrations decrease with depth. Air-sea gas exchange of dissolved oxygen is fast (months) such that the surface ocean is always close to the temperature dependent saturation concentration. (Cold water can hold more dissolved oxygen gas than warm water.) In fact in regions where significant photosynthesis takes place surface concentrations are slightly supersaturated due to oxygen production by phytoplankton. The deviation of the dissolved oxygen concentration from the saturation concentration, called Apparent Oxygen Utilization (AOU), is a measure of the total amount of organic matter oxidation of a water parcel since it has left the surface. Oxygen concentrations are low (AOU is high) in waters that have been isolated from the surface for a long time and in which lots of organic matter has remineralized such as the subsurface waters in the North Pacific. In some parts of the ocean oxygen concentrations can become close to zero (hypoxic), which can be lethal for many animals such as fish, crab or starfish. Oxygen concentrations are high in waters that have recently been at the surface such as North Atlantic Deep Water.

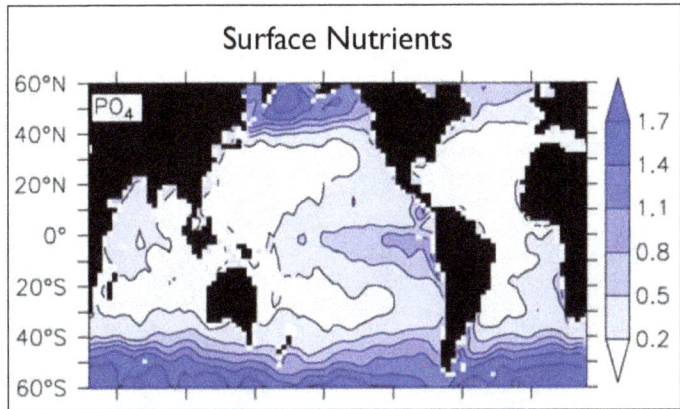

Phosphate concentrations (mmol/m³) of surface waters.

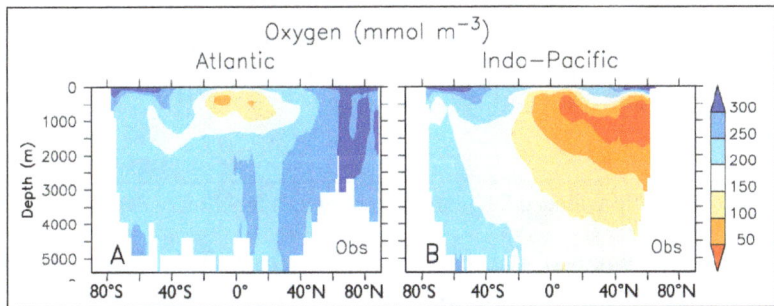

Zonally averaged dissolved oxygen concentrations.

Air sea gas exchange of carbon depends on the difference in partial pressures between the surface ocean mixed layer and the atmosphere:

$q = -K\left(\left|\vec{v}\right|, T, S\right)\left[\left(pCO_2\right)_{atm} - \left(pCO_2\right)_{ml}\right]$. The partial pressure of CO_2 in the mixed layer depends on the aquatic CO_2 concentration and the temperature (and salinity) dependent solubility α:

$$\left(pCO_2\right)_{ml} = \frac{\left|CO_2\right|^{ml}}{\alpha\left(T, S\right)}.$$

CO_2 reacts with sea water and forms carbonic acid, bicarbonate (HCO_3) and carbonate (CO_3) ions:

$$CO_2 + H_2O \Leftrightarrow HCO_3^- + H^+$$

$$HCO_3^- \Leftrightarrow CO_3^{2-} + H^+.$$

Total carbon, or dissolved inorganic carbon (DIC), is the sum of all three species:

$$DIC = \left[HCO_3^-\right] + \left[CO_3^{2-}\right] + \left[CO_2\right],$$

where aquatic CO_2 is only a small part (1%).

The sinking of biogenic matter is associated with the sinking of carbon. This is called the biological pump. The biological pump is responsible for about 2/3 of the surface to deep ocean gradient in dissolved inorganic carbon (DIC, Figure). The other 1/3 of the surface to deep ocean gradient is due to the increased solubility of CO_2 in cold waters, the so-called solubility pump.

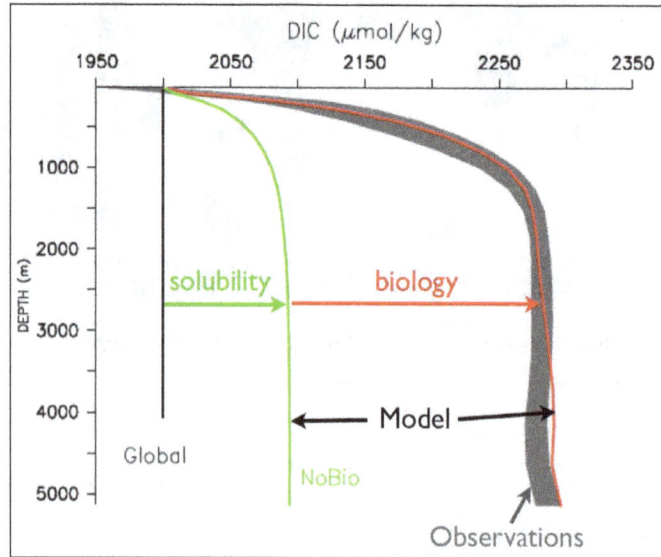

Globally averaged DIC concentrations as a function of depth.

It is useful to distinguish between the "soft tissue" pump, which is due to the sinking of particulate organic matter and the "hard tissue" pump, which is associated with the sinking of calcium carbonate (inorganic carbon). The chemical reaction equations $CO_2 + H_2O \Leftrightarrow HCO_3^- + H^+$ can be re-written in terms of (temperature dependent) equilibrium dissociation constants K_1 and K_2:

$$\left[H^+\right]\left[HCO_3^-\right] = K_1\left[CO_2\right]^{ml}$$
$$\left[H^+\right]\left[CO_3^{2-}\right] = K_2\left[HCO_3^-\right]$$

from which we can calculate the concentration of aquatic CO_2:

$$\left[CO_2\right]^{ml} = \frac{K_2\left[HCO_3^-\right]^2}{K_1\left[CO_3^{2-}\right]}.$$

Production of calcium carbonate removes carbonate ions (CO_3) thereby increasing CO_2. Thus, whereas the production of organic matter reduces surface ocean CO2 and hence atmospheric CO_2, calcium carbonate production increases surface ocean and atmospheric CO_2. Calcium carbonate is produced mainly by coccolithophorids (phytoplankton), foraminifera (zooplankton), pteropods (zooplankton) and corals. Coccolithophorids and foraminifera produce the mineral form of calcite, whereas pteropods and corals produce aragonite. Calcite and aragonite have slightly different properties, e.g. with respect to the dependence of saturation values on pH.

In addition to DIC ocean carbon cycle models must keep track of the charges, which is done by solving an equation for alkalinity (ALK):

$$ALK = \left[HCO_3^-\right] + 2\left[CO_3^{2-}\right].$$

Alkalinity is sensitive to the calcium carbonate pump. Calcium carbonate dissolves deeper (e-folding depth of 3-4 km) in the water column than organic matter (e-folding depth of 0.3-0.4 km) as shown in the deeper maximum compared with nutrients and DIC. The production and export from the euphotic zone of calcium carbonate pump is only about 10% of that of organic matter.

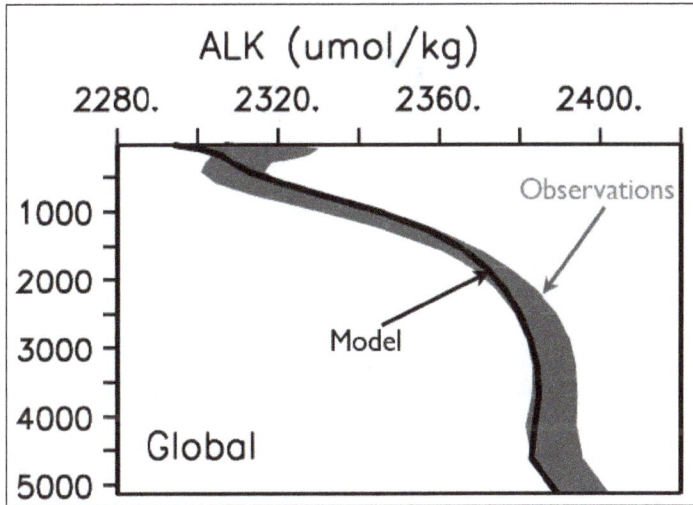

Global average alkalinity profile.

Initial attempts to model the ocean carbon cycle, such as the Ocean Carbon Model Intercomparison Project (OCMIP1) effort, were based on removing nutrients and carbon from the surface ocean and sequestering them in the deep ocean, without consideration of specific plankton functional groups. These models and most of the subsequent more complex models assume constant stoichiometric composition (C:N:P:O ratios) of the organic matter. An example of a slightly more complex model with two plankton functional groups (nitrogen fixers and other phytoplankton) and two limiting nutrients (nitrate and phosphate) one zooplankton group and a detritus compartment is shown in figure.

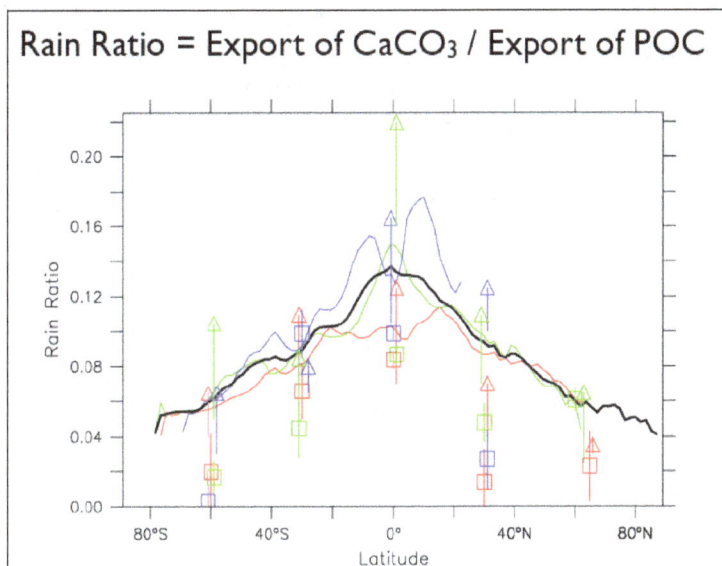

Ratio of $CaCO_3$ versus POC export out of the euphotic zone (across 120 m).

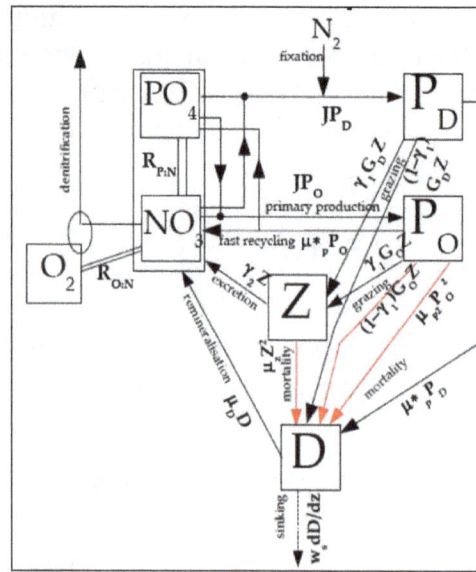

Simple model of ocean ecosystem dynamics.

Climatology

Climatology is the branch of the atmospheric sciences concerned with both the description of climate and the analysis of the causes of climatic differences and changes and their practical consequences. Climatology treats the same atmospheric processes as meteorology, but it seeks as well to identify the slower-acting influences and longer-term changes of import, including the circulation of the oceans and the small yet measurable variations in the intensity of solar radiation.

From its origins in 6th-century-BC Greek science, climatology has developed along two main lines: regional climatology and physical climatology. The first is the study of discrete and characteristic weather phenomena of a particular continental or subcontinental region. The second involves a statistical analysis of the various weather elements, principally temperature, moisture, atmospheric pressure, and wind speed, and a detailed examination of the basic relationships between such elements. Since the 1960s a third main branch, dynamic meteorology, has emerged. It deals primarily with the numerical simulation of climate and climatic change, employing models of atmospheric processes based on the fundamental equations of dynamic meteorology.

Paleoclimatology

Paleoclimatology is the study of ancient climates, prior to the widespread availability of instrumental records. Similar to the way archeologists study fossils and other physical clues to gain insight into the prehistoric past, paleoclimatologists study several different types of environmental evidence to understand what the Earth's past climate was like and why. Over the years, the Earth has kept records of its climate conditions preserved in tree rings, locked in the skeletons of tropical coral reefs, sealed in glaciers and ice caps, and buried in laminated sediments from lakes and the ocean. Scientists can use those environmental recorders to estimate past conditions, extending our understanding of climate back hundreds to millions of years.

Trees like the Giant Sequoia pictured here can grow to be hundreds or even thousands of years old, providing an important source of information about past environmental variations.

If there is one thing that the paleoclimate record shows, it's that the Earth's climate is always changing. Over the past two million years, numerous glacial periods have covered much of the high-latitude northern hemisphere landmasses in glacial ice, dropped sea level as much as 410 feet, and significantly cooled even tropical regions. In the more distant past, the Cretaceous Period (between 145.5 and 65.5 million years ago) was significantly warmer than today, with less polar ice, rising sea levels, and warm weather organisms thriving even in near-polar regions.

The study of paleoclimates has been particularly helpful in showing that the Earth's climate system can shift between dramatically different climate states in a matter of years or decades. For example, tree-ring and lake-sediment records from North America show that decadal-scale "megadroughts" occurred multiple times over the last thousand years. During these periods, persistent droughts lasted much longer than any of the droughts we have experienced over the period of instrumental records. Understanding "climate surprises" of the past is critical to avoid being surprised by abrupt climatic change in the future.

The study of past climate change also helps us understand how humans influence the Earth's climate system. The climatic record over the last thousand years clearly shows that global temperatures increased significantly in the 20th century, and that this warming was likely to have been unprecedented in the last 1,200 years. The paleoclimatic record also allows us to examine the causes of past climate change and to help unravel how much of the 20th century warming may be explained by natural causes, such as solar variability, and how much may be explained by human influences.

Paleotempestology

The study of past tropical cyclone activity by means of geological evidence and historical documentary records.

Bioclimatology

Bioclimatology is the branch of climatology that deals with the effects of the physical environment on living organisms over an extended period of time. Although Hippocrates touched on these matters 2,000 years ago in his treatise on Air, Waters, and Places, the science of bioclimatology is relatively new. It developed into a significant field of study during the 1960s owing largely to a growing concern over the deteriorating environment.

Because almost every aspect of climate and weather has some effect on living organisms, the scope of bioclimatology is almost limitless. Certain areas are emphasized more than others, however, among them studies of the influence of weather and climate on small plant organisms and insects responsible for the development of plant, animal, and human diseases; the influence of weather and climate on physiological processes in normal healthy humans and their diseases; the influence of microclimate in dwellings and urban centres on human health; and the influence of past climatic conditions on the development and distribution of plants, animals, and humans.

References

- Climate-variability, earth-science-oceanography-ocean-earth-system: science.nasa.gov, Retrieved 05 January, 2019

- The-effects-of-topography-on-the-climate-12508802: sciencing.com, Retrieved 25 May, 2019

- Climatology: britannica.com, Retrieved 23 April, 2019

- What-paleoclimatology: ncdc.noaa.gov, Retrieved 16 August, 2019

- Bioclimatology: britannica.com, Retrieved 24 July, 2019

Chapter 2

Climate of The Earth

There are different types of climate depending upon the terrain such as tropical climate, dry climate, temperate climate, continental climate, polar climate, etc. The aim of this chapter is to explore the different types of climate. These topics are crucial for a complete understanding of climatology.

Tropical Climate

The term tropical has a rather specific meaning when applied to the scientific sense of the word. An area with tropical climate is one with an average temperature of above 18 degrees Celsius (64 degrees Fahrenheit) and considerable precipitation during at least part of the year. These areas are nonarid and are generally consistent with equatorial climate conditions around the world.

Tropical Rainforests

One type of tropical area commonly spoken of is the tropical rainforest. Many tropical areas include rainforest, where temperatures range from 70 to 85 degrees Fahrenheit year-round and more than 400 inches of rain fall annually. Temperate rainforests exist in areas other than the tropics and are differentiated only by the dry season that occurs for the temperate rainforests. Tropical rainforests are home to tens of thousands of different species of animals, even though they only cover 6 to 7 percent of the world's surface.

The reason that such heavy rain is common to most of the tropics is the way in which precipitation works in these areas. Active vertical uplift and convection begin the process, which through immense amounts of sunlight cause the evaporation of a great deal of water, which in turn falls back to the earth as rain, often in heavy thunder storms nearly every day. Storms are therefore most common in midday when the sun is at its strongest.

Areas of Tropical Climate

Three very large areas conform to the definition of a tropical climate. These are the Amazon Basin in Brazil, the Congo Basin in West Africa and much to all of Indonesia. Other, less commonly known areas that are actually tropical include the savannas of Africa and semiarid areas throughout the world. Southeast Asia and Central America are two of the most well-known tropical areas, by comparison.

Dry Climate

Dry Climate is characterized by less precipitation almost all year round. The Koppen Climate Classification System has accepted dry climate as a part of the World Climatic Regions.

Dry Climate is witnessed by regular evaporation and transpiration that surpasses the level of precipitation. Dry climate is spread along the areas from 20 - 35° North and South of the equator and the continental regions of the mid-latitudes.

Dry climate is further classified into two parts namely dry arid climate and dry semiarid climate. The dry arid (desert) is a climate that envelops almost 12 % of the land surface and is popular for its xerophytic vegetation. The dry semiarid which is also known as steppe is mainly a grassland weather condition which covers 14% of the land surface.

Dry Tropical Climate is usually originated in low-latitude deserts ranging almost 18° to 28° in the hemispheres. It is situated along the center of tropics of Cancer and Capricorn, lying almost north and south of the equator.

Due to the equatorial subtropical high pressure belt and trade winds which evaporates moisture even during intense high temperature. This type of weather is common in places like southwestern United States and northern Mexico; Argentina; North Africa; South Africa; central part of Australia with recoding a temperature of 16° C.

Wet-Dry Tropical Climates is a part of the dry climate which is usually common is areas like India, Indochina, West Africa, southern Africa, South America and the north coast of Australia. This weather is characterized by wet season and dry season simultaneously. It is usually cool at the dry season and quite hot before the wet season.

Precipitation

Low and unpredictable precipitation is the primary characteristic of a dry climate. The lowest rainfall occurs in arid, or desert, areas where precipitation averages less than 35 cm (14 inches) per year, and some deserts have years with no rainfall at all. Semiarid, or steppe, regions are comprised of grasslands characterized by short grasses and scattered small bushes or sagebrush. They receive slightly more rainfall than deserts and can receive up to 70 cm (28 inches) per year. However, most semiarid regions have less than 50 cm (20 inches) of average annual precipitation.

Evaporation

Another characteristic of a dry climate is that evaporation is often greater than precipitation. This results in a climate that lacks ground moisture due to the low average rainfall and rapid evaporation of the precipitation that does fall. For example, arid regions in the Middle East average less than 20 cm of rainfall per year, but annual evaporation rates of more than 200 cm can be ten times that of precipitation. The extreme evaporation contributes to dry, coarse soils that support little plant life. Semi-arid regions with slightly more precipitation will support some grass and small bushes.

Temperature

A third common characteristic of dry climates are wide variances in seasonal and daily temperatures. Deserts are usually found in the rain shadows of mountain ranges and have hot summers, cool nights and moderate winters. However, in cold deserts, winters can be extremely frigid. In dry climates, the sun's rays are more direct, due to the lack of humidity, and this results in extreme daily temperature swings. Desert highs can approach 40 degrees Celsius (104 Fahrenheit) or more, and in some areas, winter lows can drop well below freezing.

Dry Regions

Arid and semi-arid regions together make up 26 percent of the land area of Earth, and deserts comprise 12 percent of the land. The great deserts of the world are found in the Sahara in North Africa, the Chihuahua and Sonoran deserts of Mexico and the southwestern United States, and the Gobi desert in Asia. The largest semiarid regions in the world are found in the great steppes of Russia and the short-grass plains and sagebrush areas of the North American Plains and Great Basin, as well as the Pampas of South America.

Temperate Climate

Regions of Earth can be divided into zones based on their proximity to the equator: tropical, temperate and polar. The temperate zone lies between the tropics and the polar regions and experiences a wide range of temperature and precipitation where four distinct seasons are common. Every continent except Antarctica has at least a small portion of land in the temperate zone. Regions that lie exclusively in the temperate zone include the continental U.S., most of Canada and Europe, Central Asia, southern South America and southern Australia.

Parallels of Latitude

Parallels of latitude run from 0 degrees at the equator to 90 degrees N at the north pole and 90 degrees S at the South Pole. Degrees of latitude increase as you move away from the equator and toward the poles. The temperate zone lies in the middle latitudes, in the regions of the Earth between the tropics and the polar regions. Latitude is a factor in classifying zones because it correlates to the amount of sunlight an area receives.

Climate Zones

In the early 1900s, Wladimir Koppen identified and defined climate zones of Earth: tropical, dry, temperate, continental, polar and highland. Climate zones are classified by the average temperature, amount of rainfall and type of climate they have. Latitude, elevation and the presence of nearby mountains or large bodies of water help determine climate zone because of their effects on weather patterns.

The Non-temperate Zones

The tropical zone is located at the equator and extends approximately 25 degrees to the north and south of the imaginary line. Tropical zones get more than 59 inches of rain per year. The temperature normally stays above 64 degrees Fahrenheit year-round. Dry zones receive little rainfall, and precipitation evaporates quickly. These regions are found farther away from the equator, around 20-35 degrees north and south of the equator. The polar zones, located in the high latitudes above 60-70 degrees north and south, are coldest. Temperatures generally do not exceed 50 degrees F, even in summer. Highland zones are much smaller than other zones. They are found in mountainous regions where the high elevation causes weather that is generally cool and windy and can change quickly. In the United States, highland zones are located in isolated areas of the Rocky Mountains.

Temperate Zone

In broadest sense, the temperate zone encompasses the areas of Earth that lie between the tropical zone and the polar zones. The temperate zone is sometimes called the mid-latitudes because they exist roughly between 30 degrees and 60 degrees north and south latitude. There is a greater variety in climates in the temperate zone, but many regions can be classified as either moist-continental or moist-subtropical.

Temperate Climate

The moist-subtropical climates of the temperate zone are often located near large bodies of water or far away from large mountain ranges. These regions are found at lower latitudes within the temperate zone. The winters are cool but relatively mild and summers are warm, wet and stormy. The Southeast region of the U.S. falls in this zone, as well as large portions of China, Brazil and Argentina.

The moist-continental climate zones have cold, blustery winters with plenty of snow and strong wind. Summer here is cooler than in the subtropical zones. Continental climates are located at the higher latitudes within the temperate zone and are closer to the poles than the subtropical climates. The Midwest region of the U.S, southern Canada and central Europe are classified in this climate zone.

Temperate Forests

The middle latitudes of the temperate zone experience a greater variability in temperature and precipitations than the polar or tropical zones. Although there is less biodiversity in the temperate zone than the tropical zone due to colder temperatures, 25 percent of Earth's forests reside in the temperate zone. These include both deciduous and coniferous forests. Above 50-55 degrees N latitude there are only conifer forests in the colder taiga biome. Most temperate forests are located in the Northern Hemisphere, but some temperate forests exist in New Zealand, southern Australia, South Africa, and southern Chile and Argentina. Rainforest can be found along some coastal areas in the temperate zone, such the Pacific coast in North America.

Temperate Grasslands and Deserts

Dryer regions of the temperate zone exist in the interior areas of continents, far from moist coastal air. The biomes do not get enough rainfall to support forests. Grasses thrive in areas with adequate precipitation, such as the prairies of North America, the steppes of Asia, veldt of South Africa and pampa of South America. Closer to the tropics, the temperate deserts exist in the lower latitudes of the temperate zone. Like tropical deserts, they receive less than 10 inches of annual precipitation. But unlike their tropical counterparts, experience cold temperatures in winter.

Continental Climate

Continental climates are characterized by variable weather pattern and significant variation in temperature. Continentality is the measure of the degree to which a region's climate typifies that of an interior of a large landmass. This type of climate occurs in the mid-latitudes where temperatures are not moderated by any water body such as sea or ocean and the prevailing wind blow overhead. Such regions experience colder winters and hot summers since there is no water body to keep the climate milder in winter and cooler in summer. This is because rocks and soil have a lower heat capacity compared to water and also lose heat much faster. Thus, the continental climate is relatively dry as air masses that originate from the oceans far away are lost before reaching the location. Continental climate occurs mainly in the Northern Hemisphere which has the required large landmass for the climate to develop.

General Characteristics of Continental Climate

Continental climate is characterized by a moderate amount of precipitation, mostly concentrated in the warmer months. The precipitation is derived from the frontal cyclone and conventional showers during the summer months when the maritime tropical air pushes northwards behind retreating polar front. Most areas show distinct summer precipitation maximum with only a few areas such as the mountains of the Pacific Northwest of the continent of North America and in northern Iraq, Iran, Pakistan, Afghanistan, and Central Asia show winter precipitation maximum. Part of the annual precipitation is in the form a snowfall with the snow remaining on the ground for up to a month. The mean annual precipitation is 24-47 inches, mostly in form of snow. Summer can also feature thunderstorm and cooler temperatures. Winters are generally cold but are subject to occasional mild spell resulting from the periodic incursions of tropical air. The mean temperature during winter is typically below freezing point. Spring and autumn in this climate zone vary depending on elevation and latitude. In the southern part, spring begins as early as March while in the north as late as May.

Locations with Continental Climate

Regions of the world that experience continental climate includes much of North America, Central Russia, and Siberia. Canada, Siberia, and northern states of the United States in particular exhibit large differences in wintertime and summertime average temperature of up to 40 degrees Celsius.

Vegetation of Continental Climate

The continental climate is found mainly in the inland and eastern parts of a continent. The warm summer and cold winter encourage the growth of diverse plants, from plants to perennial and ground covers. Large parts of the climate area were covered with forests before the lands were cleared for agriculture. The forests within the climate zone are divided into coniferous and deciduous. Some of the plants common in continental climate regions include silver maple, Carolina lupine, and lavender. This zone also holds different types of animals including birds and snakes.

Polar Climate

A polar climate is a type of climate in which temperatures average less than 50 °F each month of year, and therefore warm summers are not experienced. The typical temperature of polar nights is even lower than the 50 °F average, and some polar climate regions are even colder, with average temperatures of less than 0 °F. Regions that experience polar climates cover at least 20% of the Earth's surface and are situated at higher latitudes, especially near the North and the South Poles. However, no clear boundary exists to mark the location of polar climate regions. The lowest temperature ever recorded on Earth was −128.6 °F, measured at the Vostok Station in Antarctica, which is a region with a polar climate.

Characteristics of a Polar Climate

Temperature

The most apparent characteristic of a polar climate is an average monthly temperature that does not exceed 50 °F. However, some regions have much lower temperatures that never go beyond the freezing point, especially in the coldest places on Earth. Such areas include Antarctica, Greenland, and some parts of Europe.

Permanent Ice Sheet

Another characteristic of a polar climate is the presence of a permanent ice sheet. These ice sheets are formed because temperatures never reach a point warm enough for the ice to melt. Consequently, the ice sheets have accumulated over millions of years and have become very thick (as much as several kilometers thick). The existence of permanent ice sheets also mean that plants species that cannot survive in these harsh climates. In addition, animal life is scarce, although a few animal species, such as the polar bear, can feed from the oceans that extend to the fringe of polar climate regions. Like most animal species, humans cannot survive in the cold climate. While no permanent human settlements exist, temporary research stations are sometimes established in polar climates.

Precipitation

Polar climates also tends to be extremely dry since the descending cold air does not have a significant amount of moisture. Consequently, no rain clouds are formed. Some areas in polar regions receive an annual rainfall of fewer than 10 inches. If these areas were not covered with ice, they would be as dry as some of the Earth's hottest and driest deserts. The coldest place on Earth, Vostok, has an annual precipitation of only 6.5 inches. Additionally, since the ice sheets never melt, the ground is perennially dry. Any precipitation that falls comes in the form of small ice crystals or snow.

Distinction from Tundra Climate

Polar climates have some similarities to tundra climates. However, the two climates also exhibit certain differences. For example, tundra climates usually have a month in which the average temperature rises beyond the freezing point, while this does not occur in polar climates. The hotter month allows the ice in tundra climates to melt, which enable plants and animals to survive.

References

- Meaning-tropical-climate-8722483: sciencing.com, Retrieved 16 April, 2019

- Dry-climate: mapsofworld.com, Retrieved 14 August, 2019

- Characteristics-dry-climate-4878: sciencing.com, Retrieved 23 August, 2019

- Temperate-zones-located-5882122: sciencing.com, Retrieved 02 June, 2019

- What-is-the-continental-climate: worldatlas.com, Retrieved 25 April, 2019

- What-are-the-features-of-a-polar-climate: worldatlas.com, Retrieved 03 March, 2019

Chapter 3

Variables Affecting Climate

The primary factors affecting the climate include humidity, atmospheric circulation, ocean and wind currents, atmospheric pressure, winds, precipitation, temperature, geography, latitude, neutral phase, etc. This chapter closely examines these phenomena of climatology to provide an extensive understanding of the subject.

Humidity

The climate system of the Earth is ever changing across all space and time scales. Evidence for past changes arises from "proxies" such as ice cores and geological records, and for more recent times from tree rings, coral growth, and historical documentary records. Only over the last two Centuries have we been actively measuring the atmosphere.

Humidity, both relative and absolute, is potentially a very insightful tool for climate research. To constrain the earth's near-surface energy budget the concept of moist enthalpy must be understood:

$$H = C_p T + Lq$$

where H is moist static energy (or moist enthalpy) (in J kg^{-1}), C_p is the specific heat capacity of air at constant pressure, T is air temperature (in K), L is the latent heat of phase change of water vapour and q is specific humidity of air (in kg kg^{-1}). Hence, to warm a parcel of air when relative humidity is conserved (q increases with T) requires considerably more energy than if the specific humidity remains constant.

Climate model interpretations of future and past climates have generally converged upon a near-constant relative humidity over time, thus requiring more energy per 1 K rise in global temperature than implied by the temperature change alone assuming dry processes. Whether relative humidity is conserved in reality, in addition to being essential for climate model validation, has implications for how temperature changes may occur aloft. Any increases in surface water vapour (absolute humidity) will lead to greater warming aloft due to latent heating effects upon condensation. Furthermore, any changes in surface absolute humidity have implications for upper-tropospheric water vapour content, where it plays a significant role in the global radiation budget as a greenhouse gas.

Detection and attribution studies have previously considered temperature, and to a lesser extent precipitation and pressure changes, as diagnostics. Specific humidity too has potential use here with likely favourable signal-to-noise properties in warm, water abundant regions according to Clausius-Clapeyron theory. Showing consistent evidence for anthropogenic influences across a broad range of climate variables increases our confidence in the reality of human-induced global warming.

In addition, humidity has important implications for Climate Impact studies including human heat stress. There is much potential for combining historical humidity, temperature and epidemiological records with forecast capabilities to provide improved human health warnings, hospital demand forecasts etc., over timescales from days to decades.

To date, efforts to collate records of surface water vapour to form climate records have: been limited to small regions; considered only land observations; or made no attempt to ensure station homogeneity. Therefore, this thesis aims to create a truly global homogenous humidity dataset suitable for use in climate studies. Alongside efforts underway elsewhere to create and maintain upper-air humidity datasets from radiosondes and satellites, this provides potential to constrain our understanding of recent changes in both temperature and humidity throughout the troposphere.

Role of Water Vapour in the Atmosphere

Water vapour plays a key role in determining the dynamical and radiative properties of the climate system. Water vapour and its transport around the atmosphere is a fundamental component of the hydrological cycle. It also modifies the radiative balance, being a naturally occurring greenhouse gas. The Clausius-Clapeyron relation yields exponential increases in the atmosphere's water holding capacity with increasing T at approximately 7 % K^{-1}. For rising T and in the presence of unlimited water supplies (e.g. over the oceans) it can be expected that actual moisture content (i.e. specific humidity (q)) will also increase, thus maintaining a reasonably constant relative humidity (RH). Where moisture is limiting (e.g. over many land areas), q should increase less thereby allowing temperatures to increase by higher amounts and RH to decrease. It is generally assumed that RH distributions remain constant in the atmosphere over long time scales and this has been an emergent constraint on which the climate models have converged at all latitudes. However, this premise remains unproven in the observational record as to date no truly climate-quality global humidity dataset has been available.

The hydrological cycle is thought to be enhanced both in terms of weather system and precipitation intensity as higher air temperatures enable greater take up and transfer of water vapour (latent heat) from the surface to the upper atmosphere. Latent heat transfer is a major driver for atmospheric dynamics, including: the formation and propagation of mesoscale and synoptic scale weather systems; atmospheric circulation; and flooding and drought events.

Atmospheric circulation is mainly forced by latent heat release in the tropics and radiative cooling in the polar regions, giving us the more predictable modes of climate (air mass formation regions, seasonal weather characteristics, ENSO (El Niño Southern Oscillation) etc.) that regionally people have learned to live with or even come to depend upon. Exactly how these climate features might be affected by changes in surface humidity is an important question. The record 2005 North Atlantic hurricane season, in terms of intensity and heavy rainfall, has been linked by some to higher sea-surface temperature (SST) and associated increases in water vapour where resulting latent heat release is a driver for hurricanes.

Water vapour affects the earth's energy budget in four main ways. Firstly, water vapour stores energy in the form of latent heat. This is released into the atmosphere during condensation and precipitation. Secondly, as a gas it affects the energy budget through long-wave radiation effects. Indeed, it is by far the largest and the most significant of the greenhouse gases, which collectively facilitate a positive feedback mechanism which at present maintains the Earth's energy budget and cycle in a habitable state. Without these greenhouse gases, our planet's equilibrium T would lie around −18 °C rather than 14 °C (global mean surface T). Thirdly, water vapour is the source for clouds which have significant and complex radiative properties depending on their height, optical properties and latitude. Finally, the amount of moisture in the air determines the energy required to change the T of that air.

Water vapour as a greenhouse gas is part of a major positive feedback loop, increasing climate sensitivity by a factor of two. Notably, this long-wave radiation trapping is at its maximum in the mid- to upper-troposphere, despite the vertical water vapour profile which is greatest at the surface reducing rapidly with height. As water vapour can be transported vertically through convection and subsidence, and horizontally by atmospheric circulation, changes in surface absolute moisture can effect changes in moisture aloft.

An increase in atmospheric moisture provides more condensate for cloud formation. However, this and cloud development depends on many other factors such as atmospheric T, RH, stability, circulation and availability of condensation nuclei. Observed changes have been found in cloud amount, height and optical properties such as depth, liquid water content and opacity which will in turn affect radiation and thus impact on climate. These can effect a net cooling or a net warming through interaction with both short-wave and long-wave radiation. This depends on latitude, altitude and optical properties. Additionally, the quantity of atmospheric water vapour directly interacts with radiation prior to cloud interaction thus further complicating cloud feedbacks. The presentday net radiative effect of clouds is thought to be close to zero. However, cloud feedback depends on so many factors that it is uncertain as to whether climate change may impact short-wave and long-wave radiation differentially. As such, cloud feedbacks are the key source of uncertainty in climate models.

Of more direct societal interest, surface humidity has a compound effect on human comfort in terms of heat stress. High humidity in terms of RH, inhibits evaporation, making cooling by perspiration less effective contributing to higher heat-stress and potential mortality than would otherwise be expected. However, low RH too can be a source of heat stress, particularly through dryness enhancing the effect of air pollution. Generally, temperatures and humidities outside of the usually accustomed range pose a human health threat. A change in amount or distribution of surface water vapour can thus be linked to direct human health impacts.

Surface humidity may be affected by factors other than rising temperatures. These include: changes in atmospheric circulation (possibly brought about by changes in climate); changes in land-use including irrigation and reservoirs; and increasing airtraffic although this is essentially an upper-tropospheric effect.

Clearly, water vapour has played and will continue to play a key role in our changing climate as: a much affected variable; a major agent of change; and a human health issue. However, more research is urgently needed to quantify and understand recent changes, their causes, and their impacts fully.

Measuring Surface Humidity

Station observed humidity is commonly measured as one of: wet-bulb temperature (T_w); dewpoint temperature (T_{dw}); or RH. Other humidity variables can then be calculated through empirically based conversions from any of these observed parameters with the inclusion of pressure (P) and air T (dry-bulb temperatures) where necessary. There are a wide variety of instruments, collectively called hygrometers, available to measure each of the above variables.

The most commonly reported surface measure, T_w, is usually obtained using a psychrometer which contains both a dry-bulb and wet-bulb thermometer. Under suitably aspirated conditions, the contact of a hydrated wick (by means of a reservoir) around the wet-bulb thermometer causes evaporative cooling of the wet-bulb relative to the drybulb. The quantity of this depression relative to the dry-bulb temperature is directly related to the level of saturation of the air. Aspiration of the thermometer may be done: manually, by whirling the psychrometer through the air (Sling Psychrometer); mechanically, by means of a fan (Assman Psychrometer); or naturally, by situating the psychrometer in a ventilated box (often a Stevenson Screen) with adequate exposure to allow air-flow though.

Instruments to measure RH most commonly use either capacitance or resistance of an electrical current. For example, the Dewcel calculates RH from changes in the conductivity of lithium chloride as it absorbs/evaporates moisture from/to the surrounding air, requiring adequate ventilation. It can also measure T_{dw}. Chilled mirror hygrometers measure T_{dw} directly by cooling the mirror to the temperature at which moisture forms.

All instruments have potential for error. Wet-bulb thermometers can be affected by: both under- and over-ventilation depending on the Screen location; the presence of the observer which can cause positive errors in both T and T_w; and by heat conduction from dry parts of the thermometer depending on the stem length. In sub-zero temperatures where an ice-bulb calculation is used to convert to other humidity variables, the wick around the bulb may not actually be frozen and so small positive errors may occur. Furthermore, the Stevenson Screen or wick around the wet-bulb may freeze preventing ventilation of the wet-bulb (icing). The wet-bulb reservoir may freeze or in warm conditions evaporate completely, causing the wick around the wet-bulb thermometer to dry out. Icing and reservoir freezing were found to be particular problems for automatic stations in Canada, if instruments were not checked regularly. By implication this is likely a problem at other high latitude stations. Reservoir evaporation is a common problem even in temperate conditions. The 2003-2004 Global Climate Observing System (GCOS) plastic screen trial at three British stations had to discount 13 % of psychrometer measurements because the wet-bulb had dried out. All of these problems, in effect, inhibit evaporation, and thus inhibit depression of Tw relative to T giving erroneously high (frequently 100%) RH recordings and a moist bias to the data. Automated stations, which are increasingly common, are especially prone to such problems where stations are unmanned for long periods.

Retarded response of RH sensors, which increases as T decreases is also a problem, as is potential dripping of condensation down the sensor probe. The latter can be avoided by situating the RH sensor with the probe pointing skywards, as recommended by the manufacturer. However, to keep the sensor close to the dry-bulb thermometer and avoid the effect of any temperature gradient within the Screen, it is common practice to place the sensor technically upside down.

Further sources of error occur when the humidity record is physically observed as one variable but converted to and reported as another. The algorithms chosen for such conversions are not standardised and vary often between stations and even within one station over time.

Physical Relationships between Humidity Variables of Interest

Atmospheric water vapour is a complex meteorological element. For atmospheric studies at the surface it can be described in numerous ways, the most common of these being: RH; e; saturated vapour pressure (e_s); q; T_{dw}; and T_w. All except q can be measured directly.

The chosen humidity variables for this project are e, q and RH. These are selected because: they represent an absolute, proportional and a relative (respectively) measure of humidity; they are highly suitable for climate studies; they are represented in climate models (q and RH); and because they are comparable with other studies both at the surface and in the upper air.

Vapour pressure is the partial pressure of water vapour as an atmospheric gas. It is measured in mb or hPa. Within the earth's atmosphere, water vapour behaves as an ideal gas, satisfying the following conditions:

$P \propto 1/V$ at constant T

$V \propto T$ at constant P

$P \propto T$ at constant V

where V is volume in m³. It can thus be described by water vapour density (ρv) in kg m⁻³ using the equation of state:

$$e = \rho_v R_v T$$

where R_v is the specific gas constant for water vapour (462 J K⁻¹ kg⁻¹).

Specific humidity is the proportion of the mass of water vapour to the total mass of moist air. It is measured in g kg⁻¹ or kg kg⁻¹ and can be described thus:

$$q = \frac{m_v}{m_v + m_d}$$

where m_v is the mass of water vapour in kg and m_d is the mass of dry air in kg. It can also be described in terms of density because the water vapour and the moist air occupy the same total volume:

$$q = \frac{\rho_v}{\rho}$$

where ρ is the density of the moist air (kg m⁻³). If each density is then replaced by the appropriate equation of state this gives:

$$q = \frac{eR}{PR_v}$$

where R is the specific gas constant for moist air (which can be substituted with the dry air value $287 \, J \, K^{-1} \, kg^{-1}$ without causing serious error). The ratio of the gas constants R/R_v is the inverse ratio of the molecular weights of each and can be substituted with 0.622 (known as ε). Thus q can be derived from e as follows:

$$q = \frac{\varepsilon e}{P}$$

where e and P are both either in mb or hPa, and q is output in $kg \, kg^{-1}$.

Importantly, e recorded at T is by definition the same as e_s at the simultaneous T_{dw} ($e_s(T_{dw})$). The saturation vapour pressure is the partial pressure of water vapour on a parcel of air should that parcel become saturated at its current T ($e_s(T)$), essentially a measure of the water holding capacity of the air. The saturation specific humidity (q_s) is the equivalent for q. From this knowledge of both the actual moisture content of the air and potential moisture content of the air (at saturation), the relative humidity, which refers to the extent of saturation of the air as a percentage, can be calculated:

$$RH = 100 \left(\frac{e_s\left(T_{dw}\right)}{e_s\left(T\right)} \right)$$

which can be rewritten:

$$RH = 100 \left(\frac{e}{e_s} \right)$$

Humidity in Climate Models

General circulation models (GCMs) or climate models have become increasingly important in climate analyses and will continue to do so for the foreseeable future. The most sophisticated incorporate a wide range of forcings including greenhouse gases, land use change, volcanoes, solar output changes and fully interactive natural and anthropogenic aerosol modelling. While GCMs are widely thought of as useful tools in climate research, their limitations should be noted. Necessarily, many sub-grid-box scale physical processes must be parameterised, especially in cases where these processes occur on scales smaller than the grid-box resolution of the model. There is considerable uncertainty in these parameterisations. Notably, cloud feedbacks and many of the critical process controlling water vapour are parameterised and are consequently areas of large uncertainty.

At least for T, coupled (atmosphere and ocean) models can now provide credible climate simulation down to sub-continental and seasonal scales. However, little work has been done to date to validate humidity in the models, with virtually no mention within the Third Assessment Report of the IPCC. This is most likely due to a lack of suitable observations with which to compare the models.

The majority of the literature relating climate models to observed or theoretical humidity changes refers to the positive feedback mechanism of water vapour, and RH in the free troposphere. The water vapour feedback occurs in models because increasing temperatures lead to increases in atmospheric water vapour content which as a greenhouse gas leads to further warming. Huang et al. compared tropical midand upper-tropospheric humidity model output of GFDL (Geophysical Fluid Dynamics Laboratory) AM2 with HIRS radiance measurements and found good agreement. Constant RH over large spatial and temporal scales is a feature exhibited in most climate models. It originates from physical processes within the model as opposed to initial assumptions. This is even the case in low latitudes where this was not thought so plausible. These findings support near constant RH as a robust constraint on atmospheric humidity such that q could be expected to increase exponentially with T following Clausius-Clapeyron theory. Indeed, model projections have shown a possible doubling of atmospheric water vapour by 2100 resulting from increasing greenhouse gas forcings on climate, with the increasing water vapour, through radiation, leading to further increases in temperature.

Observed trends in surface q (1976 to 1999) were found broadly comparable to those from the coupled Parallel Climate Model (PCM) as were q-T correlations. In the model, the correlations were slightly higher and inter-annual variability underestimated. It should be noted, however, that a good ability to simulate past climates does not necessarily mean that responses to future perturbations remain plausible.

Atmospheric humidity means the amount of water vapour or moisture in the air, is another leading climatic element, as is precipitation. All forms of precipitation, including drizzle, rain, snow, ice crystals, and hail, are produced as a result of the condensation of atmospheric moisture that forms clouds in which some of the particles, by growth and aggregation, attain sufficient size to fall from the clouds and reach the ground.

At 30 °C (86 °F), 4 percent of the volume of the air may be occupied by water molecules, but, where the air is colder than −40 °C (−40 °F), less than one-fifth of 1 percent of the air molecules can be water. Although the water vapour content may vary from one air parcel to another, these limits can be set because vapour capacity is determined by temperature. Temperature has profound effects upon some of the indexes of humidity, regardless of the presence or absence of vapour.

The connection between an effect of humidity and an index of humidity requires simultaneous introduction of effects and indexes. Vapour in the air is a determinant of weather, because it first absorbs the thermal radiation that leaves and cools Earth's surface and then emits thermal radiation that warms the planet. Calculation of absorption and emission requires an index of the mass of water in a volume of air. Vapour also affects the weather because, as indicated above, it condenses into clouds and falls as rain or other forms of precipitation. Tracing the moisture-bearing air masses requires a humidity index that changes only when water is removed or added.

Humidity Indexes

Absolute Humidity

Absolute humidity is the vapour concentration or density in the air. If m_v is the mass of vapour in a volume of air, then absolute humidity d_v is simply $d_v = m_v / V$, in which V is the volume and d_v is expressed in grams per cubic metre. This index indicates how much vapour a beam of radiation must pass through. The ultimate standard in humidity measurement is made by weighing the amount of water gained by an absorber when a known volume of air passes through it; this measures absolute humidity, which may vary from 0 gram per cubic metre in dry air to 30 grams per cubic metre (0.03 ounce per cubic foot) when the vapour is saturated at 30 °C. The d_v of a parcel of air changes, however, with temperature or pressure even though no water is added or removed, because, as the gas equation states, the volume V increases with the absolute, or Kelvin, temperature and decreases with the pressure.

Specific Humidity

The meteorologist requires an index of humidity that does not change with pressure or temperature. A property of this sort will identify an air mass when it is cooled or when it rises to lower pressures aloft without losing or gaining water vapour. Because all the gases will expand equally, the ratios of the weight of water to the weight of dry air, or the dry air plus vapour, will be conserved during such changes and will continue identifying the air mass.

The mixing ratio r is the dimensionless ratio $r = m_v / m_a$, where m_a is the mass of dry air, and the specific humidity q is another dimensionless ratio $q = m_v / (m_a + m_v)$. Because m_v is less than 3 percent of m_a at normal pressure and temperatures cooler than 30 °C, r and q are practically equal. These indexes are usually expressed in grams per kilogram because they are so small; the values range from 0 grams per kilogram in dry air to 28 grams per kilogram in saturated air at 30 °C. Absolute and specific humidity indexes have specialized uses, so they are not familiar to most people.

Relative Humidity

Relative humidity (U) is so commonly used that a statement of humidity, without a qualifying adjective, can be assumed to be relative humidity. U can be defined, then, in terms of the mixing ratio r that was introduced above. $U = 100r / r_w$, which is a dimensionless percentage. The divisor r_w is the saturation mixing ratio, or the vapour capacity. Relative humidity is therefore the water vapour content of the air relative to its content at saturation. Because the saturation mixing ratio is a function of pressure, and especially of temperature, the relative humidity is a combined index of the environment that reflects more than water content. In many climates the relative humidity rises to about 100 percent at dawn and falls to 50 percent by noon. A relative humidity of 50 percent may reflect many different quantities of vapour per volume of air or gram of air, and it will not likely be proportional to evaporation.

An understanding of relative humidity thus requires a knowledge of saturated vapour,. At this point, however, the relation between U and the absorption and retention of water from the air must be considered. Small pores retain water more strongly than large pores; thus, when a porous material is set out in the air, all pores larger than a certain size (which can be calculated from the relative humidity of the air) are dried out.

The water content of a porous material at air temperature is fairly well indicated by the relative humidity. The complexity of actual pore sizes and the viscosity of the water passing through them makes the relation between U and moisture in the porous material imperfect and slowly achieved. The great suction also strains the walls of the capillaries, and the consequent shrinkage is used to measure relative humidity.

The absorption of water by salt solutions is also related to relative humidity without much effect of temperature. The air above water saturated with sodium chloride is maintained at 75 to 76 percent relative humidity at a temperature between 0 and 40 °C (32 and 104 °F).

In effect, relative humidity is a widely used environmental indicator, but U does respond drastically to changes in temperatures as well as moisture, a response caused by the effect of temperature upon the divisor r_w in U.

Relation between Temperature and Humidity

Tables that show the effect of temperature upon the saturation mixing ratio r_w are readily available. Humidity of the air at saturation is expressed more commonly, however, as vapour pressure. Thus, it is necessary to understand vapour pressure and in particular the gaseous nature of water vapour.

The pressure of the water vapour, which contributes to the pressure of the atmosphere, can be calculated from the absolute humidity d_v by the gas equation:

$$e = \frac{m_v}{V}\frac{RT}{M_w} = d_v \frac{RT}{M_w},$$

in which R is the gas constant, T the absolute temperature, M_w the molecular weight of water, and e the water vapour pressure in millibars (mb).

Relative humidity can be defined as the ratio of the vapour pressure of a sample of air to the saturation pressure at the existing temperature. Further, the capacity for vapour and the effect of temperature can now be presented in the usual terms of saturation vapour pressure.

Within a pool of liquid water, some molecules are continually escaping from the liquid into the space above, while more and more vapour molecules return to the liquid as the concentration of vapour rises. Finally, equal numbers are escaping and returning; the vapour is then saturated, and its pressure is known as the saturation vapour pressure, e_w. If the liquid and vapour are warmed, relatively more molecules escape than return, and e_w rises. There is also a saturation pressure with respect to ice. The vapour pressure curve of water has the same form as the curves for many other substances. Its location is fixed, however, by the boiling point of 100 °C (212 °F), where the saturation vapour pressure of water vapour is 1,013 mb (1 standard atmosphere), the standard pressure of the atmosphere at sea level. The decrease of the boiling point with altitude can be calculated. For example, the saturation vapour pressure at 40 °C (104 °F) is 74 mb (0.07 standard atmosphere), and the standard atmospheric pressure near 18,000 metres (59,000 feet) above sea level is also 74 mb; thus, it is where water boils at 40° C.

The everyday response of relative humidity to temperature can be easily explained. On a summer morning, the temperature might be 15 °C (59 °F) and the relative humidity 100 percent. The vapour pressure would be 17 mb (0.02 standard atmosphere) and the mixing ratio about 11 parts per thousand (11 grams of water per kilogram of air by weight). During the day the air could warm to 25 °C (77 °F), while evaporation could add little water. At 25 °C the saturation pressure is fully 32 mb (0.03 standard atmosphere). If, however, little water has been added to the air, its vapour pressure will still be about 17 mb. Thus, with no change in vapour content, the relative humidity of the air has fallen from 100 to only 53 percent, illustrating why relative humidity does not identify air masses.

The meaning of dew-point temperature can be illustrated by a sample of air with a vapour pressure of 17 mb. If an object at 15 °C is brought into the air, dew will form on the object. Hence, 15 °C is the dew-point temperature of the air—i.e., the temperature at which the vapour present in a sample of air would just cause saturation or the temperature whose saturation vapour pressure equals the present vapour pressure in a sample of air. Below freezing, this index is called the frost point. There is a one-to-one correspondence between vapour pressure and dew point. The dew point has the virtue of being easily interpreted because it is the temperature at which a blade of grass or a pane of glass will become wet with dew from the air. Ideally, it is also the temperature of fog or cloud formation.

The clear meaning of dew point suggests a means of measuring humidity. A dew-point hygrometer was invented in 1751. For this instrument, cold water was added to water in a vessel until dew formed on the vessel, and the temperature of the vessel, the dew point, provided a direct index of humidity. The greatest use of the condensation hygrometer has been to measure humidity in the upper atmosphere, where a vapour pressure of less than a thousandth millibar makes other means impractical.

Another index of humidity, the saturation deficit, can also be understood by considering air with a vapour pressure of 17 mb. At 25 °C the air has (31 – 17), or 14, mb less vapour pressure than saturated vapour at the same temperature; that is, the saturation deficit is 14 mb (0.01 standard atmosphere).

The saturation deficit has the particular utility of being proportional to the evaporation capability of the air. The saturation deficit can be expressed as:

$$e_w - e = e_w \left(1 - \frac{U}{100}\right),$$

and because the saturation vapour pressure e_w rises with rising temperature, the same relative humidity will correspond to a greater saturation deficit and evaporation at warm temperatures.

Humidity and Climate

The small amount of water in atmospheric vapour, relative to water on Earth, belies its importance. Compared with one unit of water in the air, the seas contain at least 100,000 units, the great glaciers 1,500, the porous earth nearly 200, and the rivers and lakes 4 or 5. The effectiveness of the vapour in the air is magnified, however, by its role in transferring water from sea to land by the media of clouds and precipitation and that in absorbing radiation.

The vapour in the air is the invisible conductor that carries water from sea to land, making terrestrial life possible. Fresh water is distilled from the salt seas and carried over land by the wind. Water evaporates from vegetation, and rain falls on the sea too, but the sea is the bigger source, and rain that falls on land is most important to humans. The invisible vapour becomes visible near the surface as fog when the air cools to the dew point. The usual nocturnal cooling will produce fog patches in cool valleys. Or the vapour may move as a tropical air mass over cold land or sea, causing widespread and persistent fog, such as occurs over the Grand Banks off Newfoundland. The delivery of water by means of fog or dew is slight, however.

When air is lifted, it is carried to a region of lower pressure, where it will expand and cool. It may rise up a mountain slope or over the front of a cooler, denser air mass. If condensation nuclei are absent, the dew point may be exceeded by the cooling air, and the water vapour becomes supersaturated. If nuclei are present or if the temperature is very low, however, cloud droplets or ice crystals form, and the vapour is no longer in the invisible guise of atmospheric humidity.

The invisible vapour has another climatic role—namely, absorbing and emitting radiation. The temperature of Earth and its daily variation are determined by the balance between incoming and outgoing radiation. The wavelength of the incoming radiation from the Sun is mostly shorter than 3 μm (0.0001 inch). It is scarcely absorbed by water vapour, and its receipt depends largely upon cloud cover. The radiation exchanged between the atmosphere and Earth's surface and the eventual loss to space is in the form of long waves. These long waves are strongly absorbed in the 3- to 8.5-μm band and in the greater than 11-μm range, where vapour is either partly or wholly opaque. much of the radiation that is absorbed in the atmosphere is emitted back to Earth, and the surface receipt of long waves, primarily from water vapour and carbon dioxide in the atmosphere, is slightly more than twice the direct receipt of solar radiation at the surface. Thus, the invisible vapour in the atmosphere combines with clouds and the advection (horizontal movement) of air from different regions to control the surface temperature.

The world distribution of humidity can be portrayed for different uses by different indexes. To appraise the quantity of water carried by the entire atmosphere, the moisture in an air column above a given point on Earth is expressed as a depth of liquid water. It varies from 0.5 mm (0.02 inch) over the Himalayas and 2 mm (0.08 inch) over the poles in winter to 8 mm (0.3 inch) over the Sahara, 54 mm (2 inches) in the Amazon region, and 64 mm (2.5 inches) over India during the wet season. During summer the air over the United States transports 16 mm (0.6 inch) of water vapour over the Great Basin and 45 mm (1.8 inches) over Florida.

The humidity of the surface air may be mapped as vapour pressure, but a map of this variable looks much like that of temperature. Warm places are moist, and cool ones are dry; even in deserts the vapour pressure is normally 13 mb (0.01 standard atmosphere), whereas over the northern seas it is only about 4 mb (0.004 standard atmosphere). Certainly the moisture in materials in two such areas will be just the opposite, so relative humidity is a more widely useful index.

Average Relative Humidity

The average relative humidity for July reveals the humidity provinces of the Northern Hemisphere when aridity is at a maximum. At other times the relative humidity generally will be higher. The humidities over the Southern Hemisphere in July indicate the humidities that comparable regions in the Northern Hemisphere will attain in January, just as July in the Northern Hemisphere suggests the humidities in the Southern Hemisphere during January. A contrast is provided by comparing a humid cool coast to a desert. The midday humidity on the Oregon coast, for example, falls only to 80 percent, whereas in the Nevada desert it falls to 20 percent. At night the contrast is less, with averages being over 90 and about 50 percent, respectively.

Although the dramatic regular decrease of relative humidity from dawn to midday has been attributed largely to warming rather than declining vapour content, the content does vary regularly. In humid environments, daytime evaporation increases the water vapour content of the air, and the mixing ratio, which may be about 12 grams per kilogram, rises by 1 or 2 grams per kilogram in temperate places and may attain 16 grams per kilogram in a tropical rainforest. In arid environments, however, little evaporation moistens the air, and daytime turbulence tends to bring down dry air; this decreases the mixing ratio by as much as 2 grams per kilogram.

Humidity also varies regularly with altitude. On the average, fully half the water in the atmosphere lies below 0.25 km (about 0.2 mile), and satellite observations over the United States in April revealed 1 mm (0.04 inch) or less of water in all the air above 6 km (4 miles). A cross section of the atmosphere along 75° W longitude shows a decrease in humidity with height and toward the poles. The mixing ratio is 16 grams per kilogram just north of the Equator, but it decreases to 1 gram per kilogram at 50° N latitude or 8 km (5 miles) above the Equator. The transparent air surrounding mountains in fair weather is very dry indeed.

Closer to the ground, the water vapour content also changes with height in a regular pattern. When water vapour is condensing on Earth's surface at night, the content is greater aloft than at the ground; during the day the content is, in most cases, less aloft than at the ground because of evaporation.

Evaporation and Humidity

Evaporation, mostly from the sea and from vegetation, replenishes the humidity of the air. It is the change of liquid water into a gaseous state, but it may be analyzed as diffusion. The rate of diffusion, or evaporation, will be proportional to the difference between the pressure of the water vapour in the free air and the vapour that is next to, and saturated by, the evaporating liquid. If the liquid and air have the same temperature, evaporation is proportional to the saturation deficit. It is also proportional to the conductivity of the medium between the evaporator and the free air. If the evaporator is open water, the conductivity will increase with ventilation. But if the evaporator is a leaf, the diffusing water must pass through the still air within the minute pores between the water inside and the dry air outside. In this case the porosity may modify the conductivity more than ventilation.

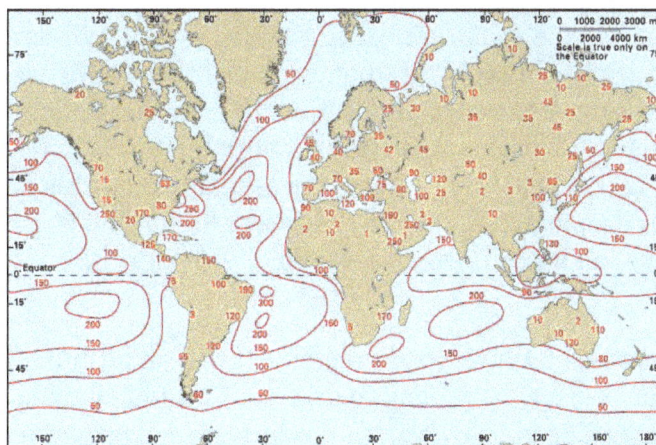

Global distribution of mean annual evaporation (in centimetres).

The temperature of the evaporator is rarely the same as the air temperature, however, because each gram of evaporation consumes about 600 calories (2,500 joules) and thus cools the evaporator. The availability of energy to heat the evaporator is therefore as important as the saturation deficit and conductivity of the air. Outdoors, some of this heat may be transferred from the surrounding air by convection, but much of it must be furnished by radiation. Evaporation is faster on sunny days than on cloudy ones not only because the sunny day may have drier air but also because the Sun warms the evaporator and thus raises the vapour pressure at the evaporator. In fact, according to the well-known Penman calculation of evaporation (an equation that considers potential evaporation as a function of humidity, wind speed, radiation, and temperature), this loss of water is essentially determined by the net radiation balance during the day.

Atmospheric Circulation

The circulation of wind in the atmosphere is driven by the rotation of the earth and the incoming energy from the sun. Wind circulates in each hemisphere in three distinct cells which help transport energy and heat from the equator to the poles. The winds are driven by the energy from the sun at the surface as warm air rises and colder air sinks.

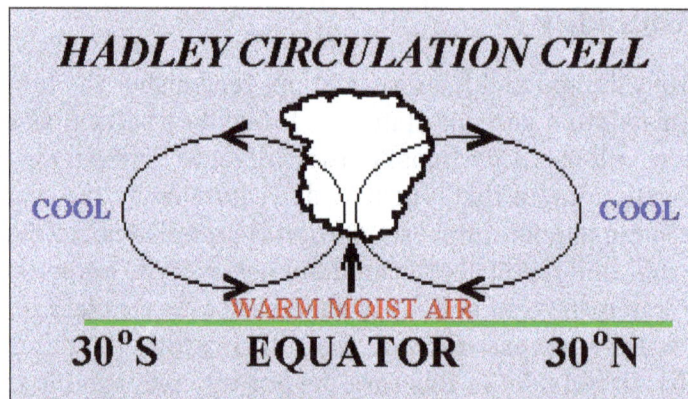

Hadley Cell circulation.

The circulation cell closest to the equator is called the Hadley cell. Winds are light at the equator because of the weak horizontal pressure gradients located there. The warm surface conditions result in locally low pressure. The warm air rises at the equator producing clouds and causing instability in the atmosphere. This instability causes thunderstorms to develop and release large amounts of latent heat. Latent heat is just energy released by the storms due to changes from water vapor to liquid water droplets as the vapor condenses in the clouds, causing the surrounding air to become more warm and moist, which essentially provides the energy to drive the Hadley cell.

The Hadley Cell encompasses latitudes from the equator to about 30°. At this latitude surface high pressure causes the air near the ground to diverge. This forces air to come down from aloft to "fill in" for the air that is diverging away from the surface high pressure. The air flowing northward from the equator high up in the atmosphere is warm and moist compared to the air nearer the poles. This causes a strong temperature gradient between the two different air masses and a jet stream results. At the 30° latitudes, this jet is known as the subtropical jet stream which flows from west to east in both the Northern and Southern Hemispheres. Clear skies generally prevail throughout the surface high pressure, which is where many of the deserts are located in the world.

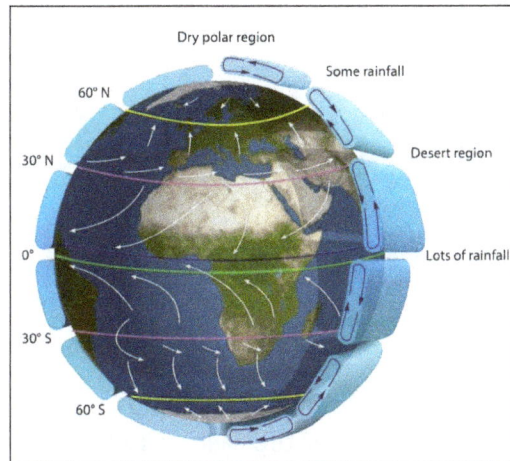

General Wind Directions.

From 30° latitude, some of the air that sinks to the surface returns to the equator to complete the Hadley Cell. This produces the northeast trade winds in the Northern Hemisphere and the southeast trades in the Southern Hemisphere. The Coriolis force impacts the direction of the wind flow. In the Northern Hemisphere, the Coriolis force turns the winds to the right. In the Southern Hemisphere, the Coriolis force turns the winds to the left.

From 30° latitude to 60° latitude, a new cell takes over known as the Ferrel Cell. This cell produces prevailing westerly winds at the surface within these latitudes. This is because some of the air sinking at 30° latitude continues traveling northward toward the poles and the Coriolis force bends it to the right (in the Northern Hemisphere). This air is still warm and at roughly 60° latitude approaches cold air moving down from the poles. With the converging air masses at the surface, the low surface pressure at 60° latitude causes air to rise and form clouds. Some of the rising warm air returns to 30° latitude to complete the Ferrel Cell.

The two air masses at 60° latitude do not mix well and form the polar front which separates the warm air from the cold air. Thus the polar front is the boundary between warm tropical air masses and the colder polar air moving from the north. (The use of the word "front" is from military terminology; it is where opposing armies clash in battle.) The polar jet stream aloft is located above the polar front and flows generally from west to east. The polar jet is strongest in the winter because of the greater temperature contrasts than during the summer. Waves along this front can pull the boundary north or south, resulting in local warm and cold fronts which affect the weather at particular locations.

Above 60° latitude, the polar cell circulates cold, polar air equatorward. The air from the poles rises at 60° latitude where the polar cell and Ferrel cell meet, and some of this air returns to the poles completing the polar cell. Because the wind flows from high to low pressure and taking into account the effects of the Coriolis force, the winds above 60° latitude are prevailing easterlies.

Walker Circulation

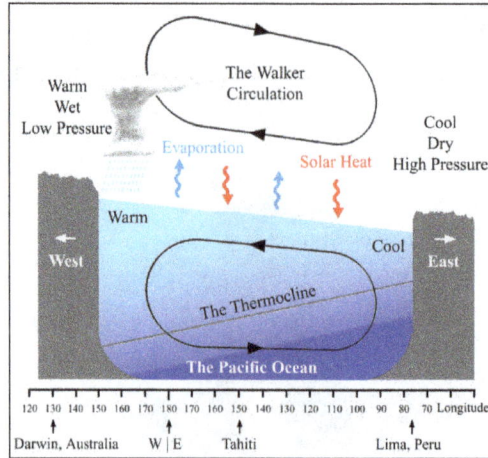

In contrast to the Hadley, Ferrel and polar circulations that run along north-south lines, the Walker circulation is an east-west circulation. Over the eastern Pacific Ocean, surface high pressure off the west coast of South America enhances the strength of the easterly trade winds found near the equator. The winds blow away from the high pressure toward lower pressure near Indonesia. Upwelling, the rising of colder water from the deep ocean to the surface, occurs in the eastern Pacific along South America near Ecuador and Peru. This cold water is especially nutrient-rich and is stocked with an abundance of large fish populations. By contrast the water in the western Pacific, near Indonesia, is relatively warm. The air over Indonesia rises because of the surface low pressure located there and forms clouds. This causes heavy precipitation to fall over the western tropical Pacific throughout the year. The air then circulates back aloft towards the region above the surface high pressure near Ecuador and this becomes the Walker circulation. The air sinks at this surface high pressure and is picked up by the strong trade winds to continue the cycle.

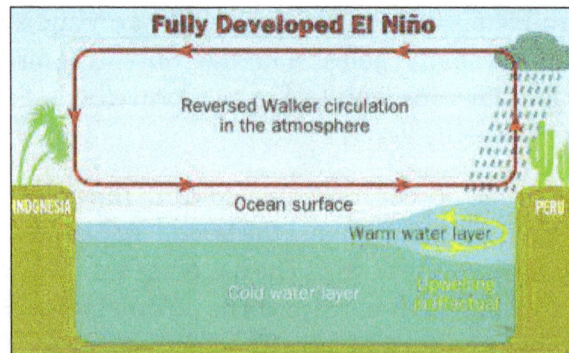

On some occasions, the Walker circulation and the trade winds weaken, allowing warmer water to "slosh back" towards the eastern tropical Pacific near South America. Consider an example of a blowing fan over a bathtub which is full of water. If the fan blows steadily, the water at the side farthest from the fan will tend to pile up downwind. If you suddenly slow the fan down, some of the water that was built up will surge back towards the fan. The warmer water will cover the areas of upwelling, cutting off the flow of nutrients to the fish and animals that live in the eastern Pacific Ocean. This warming of the eastern Pacific Ocean is known as El Niño. The warmer water will also serve as a source for warm, moist air which can aid in the development of heavy thunderstorms over the mass of warm water.

Ocean and Wind Currents

Ocean and wind currents are formed by a process known as convection. Both convection and pressure affect the flow of water and air. As air and water currents move from one area to another, they affect the general climate of the area they are moving into.

Convection

Convection is one of the major ways that heat is transferred. It occurs because hotter liquids and gasses have a tendency to rise, while colder liquids and gasses have a tendency to sink. Think of heating a pot of water on the stove. Initially, the bottom portion of the water is heated by the energy produced by the stove, but, after a while, bubbles form and rise to the surface. The bubbles are pockets of hot water rising to the surface that heat the water around them as they rise. The same thing happens on a larger scale when the sun heats the ocean and colder water sinks underneath.

Ocean Currents

Ocean currents affect the temperature by moving hot or cold water from one location to another. The Gulf Stream, for example, moves warm air from the Gulf of Mexico along the eastern coast of the U.S., and eventually to the British Isles. As the warm water travels North, it warms the water and air around it.

Air Currents

The dominant air currents that affect climate are known as prevailing winds. Prevailing winds are winds that blow in one direction more often than from other directions. Prevailing winds bring air from one type of climate to another. For example, warm winds that travel over water tend to collect moisture as they travel; the water vapor in the air will condense as it moves into colder climates, which is why temperate coastal areas often receive heavy rainfall.

Air Pressure

Another factor that affects air currents is air pressure. The higher the difference in air pressure between two areas, the stronger the winds will be. This happens because high pressure air has a tendency to move towards areas of lower pressure. Low pressured air also holds less heat than high pressured air, which is why it is generally colder at higher elevations.

Ocean Circulation

Ocean circulation is a key regulator of climate by storing and transporting heat, carbon, nutrients and freshwater all around the world. Complex and diverse mechanisms interact with one another to produce this circulation and define its properties.

Ocean circulation can be conceptually divided into two main components: a fast and energetic wind-driven surface circulation, and a slow and large density-driven circulation which dominates the deep sea.

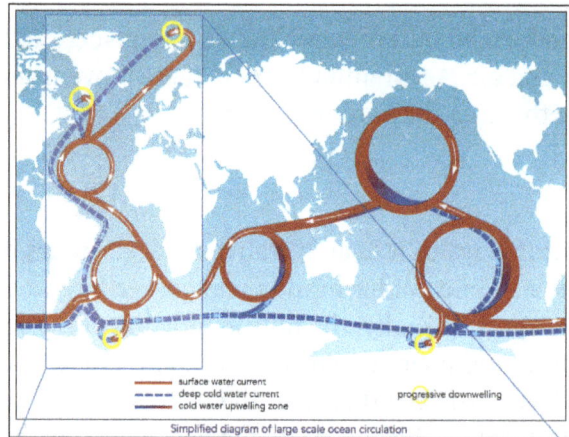

Simplified diagram of large scale ocean circulation

Wind-driven circulation is by far the most dynamic. Blowing wind produces currents at the surface of the ocean which are oriented at 90° to its direction (on its right in the Northern Hemisphere and on its left in the Southern Hemisphere) due to the Earth rotation. As a consequence, it creates zones of convergence or divergence of ocean currents at the point where they meet. Divergence of currents will create an upwelling phase (interior waters reach the surface) and convergence a downwelling phase (surface waters sink in the interior ocean), linking surface and interior waters.

The slow and deep circulation is largely driven by water density, and thus its temperature and salinity. It acts on the ocean as a whole and has a major influence on the abyssal properties where wind-driven circulation has no effect. However, this circulation is slow and generates weak currents, it is therefore more difficult to observe: a single drop of water travels 1,000 years to close the global overturning circulation.

The large-scale circulation is relatively stable on long timescales. At some very specific locations – mainly in the Northern Atlantic and around Antarctica – surface waters become denser and sink to the depths. Densification occurs due to cooling surface waters and increasing salinity, the latter as a result of the removal of freshwater and the formation of ice. Surface waters are then pulled up to replace the sinking ones. How waters up well from depths to the surface is still unclear. As stated above, zones of divergence of waters are of critical importance for these phenomena but near-sea-floor turbulence also plays a major role. These mechanisms are still poorly understood and their spatial variability remains largely unknown.

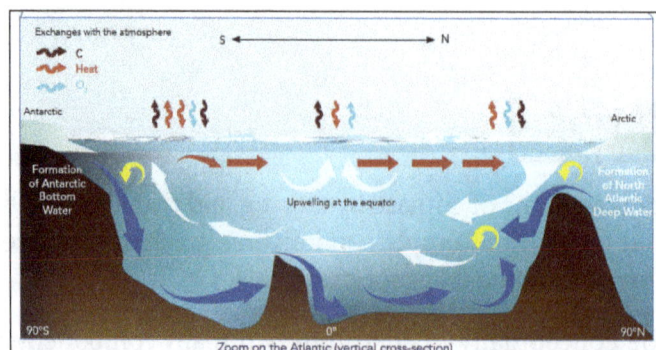

Zoom on the Atlantic (vertical cross-section)

Oceanic circulation is very sensitive to the global freshwater flux. This flux can be described as the difference between [Evaporation + Sea Ice Formation], which enhances salinity, and [Precipitation + Runoff + Ice melt], which decreases salinity. Global warming will undoubtedly lead to more ice melting in the poles and thus larger additions of freshwaters in the ocean at high latitudes. This input of freshwater, by decreasing surface water density near the poles, could limit down welling, and prevent deep waters formation, slowing down global circulation.

Such a process could have tremendous consequences for our societies. It would mean less carbon and heat uptake by the ocean and thus higher rates of both carbon and heat in the atmosphere. It could potentially accelerate global warming and enhance its negative effects.

More generally, it is important to note that interactions between oceanic circulation and climate are still poorly understood: more observations, an increased understanding and reliable numerical models of oceanic circulation are needed at different space and time scales.

Surface Currents

Ocean water moves in predictable ways along the ocean surface. Surface currents can flow for thousands of kilometers and can reach depths of hundreds of meters. These surface currents do not depend on weather; they remain unchanged even in large storms because they depend on factors that do not change. Surface currents are created by three things: global wind patterns, the rotation of the Earth, and the shape of the ocean basins.

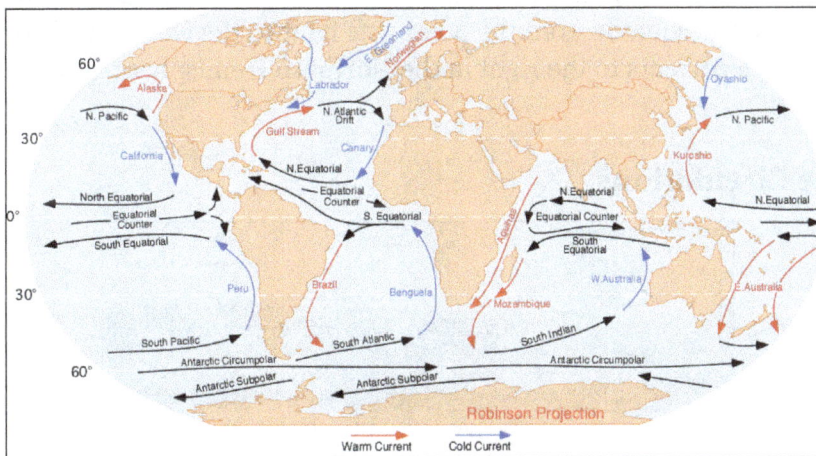

Surface ocean currents.

Surface currents are extremely important because they distribute heat around the planet and are a major factor influencing climate around the globe.

Global Wind Currents

Winds on Earth are either global or local. Global winds blow in the same directions all the time and are related to the unequal heating of Earth by the Sun, that is that more solar radiation strikes the equator than the polar regions, and the rotation of the Earth called the Coriolis effect. Water in the surface currents is pushed in the direction of the major wind belts:

- Trade winds: East to west between the equator and 30 degrees North and 30 degrees South.

- Westerlies: west to east in the middle latitudes.

- Polar easterlies: east to west between 50 degrees and 60 degrees north and south of the equator and the north and south pole.

Rotation of the Earth

Wind is not the only factor that affects ocean currents. The Coriolis effect describes how Earth's rotation steers winds and surface ocean currents. The Coriolis effect causes freely moving objects to appear to move to the right in the Northern Hemisphere and to the left in the Southern Hemisphere. The objects themselves are actually moving straight, but the Earth is rotating beneath them, so they seem to bend or curve.

An example might make the Coriolis effect easier to visualize. If an airplane flies 500 miles due north, it will not arrive at the city that was due north of it when it began its journey. Over the time it takes for the airplane to fly 500 miles, that city moved, along with the Earth it sits on. The airplane will therefore arrive at a city to the west of the original city (in the Northern Hemisphere), unless the pilot has compensated for the change. So to reach his intended destination, the pilot must also veer right while flying north.

As wind or an ocean current moves, the Earth spins underneath it. As a result, an object moving north or south along the Earth will appear to move in a curve, instead of in a straight line. Wind or water that travels toward the poles from the equator is deflected to the east, while wind or water that travels toward the equator from the poles gets bent to the west. The Coriolis effect bends the direction of surface currents to the right in the Northern Hemisphere and left in the Southern Hemisphere.

Thermohaline Circulations

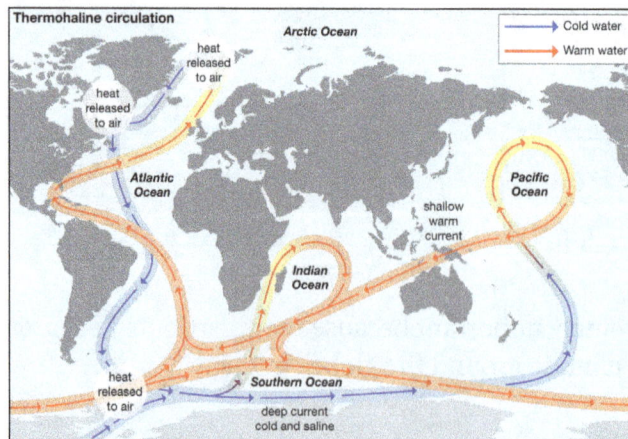

Thermohaline circulation transports and mixes the water of the oceans. In the process it transports heat, which influences regional climate patterns. The density of seawater is determined by the temperature and salinity of a volume of seawater at a particular location. The difference in density between one location and another drives the thermohaline circulation.

Thermohaline circulation is also called Global Ocean Conveyor or Great Ocean Conveyor Belt, the component of general oceanic circulation controlled by horizontal differences in temperature and salinity. It continually replaces seawater at depth with water from the surface and slowly replaces surface water elsewhere with water rising from deeper depths. Although this process is relatively slow, tremendous volumes of water are moved, which transport heat, nutrients, solids, and other materials vast distances. Thermohaline circulation also drives warmer surface waters poleward from the subtropics, which moderates the climate of Iceland and other coastal areas of Europe.

The general circulation of the oceans consists primarily of wind-driven ocean currents. These, however, are superimposed on the much more sluggish circulation driven by horizontal differences in temperature and salinity—namely, thermohaline circulation. Wind-driven circulation, which is strongest in the surface layer of the ocean, is the more vigorous of the two and is configured as large gyres that dominate an ocean region. In contrast, thermohaline circulation is much slower, with a typical speed of 1 centimetre (0.4 inch) per second, but this flow extends to the seafloor and forms circulation patterns that envelop the global ocean.

In some areas of the ocean, generally during the winter season, cooling or net evaporation causes surface water to become dense enough to sink. Convection penetrates to a level where the density of the sinking water matches that of the surrounding water. It then spreads slowly into the rest of the ocean. Other water must replace the surface water that sinks. This sets up the thermohaline circulation. The basic thermohaline circulation is one of sinking of cold water in the polar regions, chiefly in the northern North Atlantic and near Antarctica. These dense water masses spread into the full extent of the ocean and gradually upwell to feed a slow return flow to the sinking regions.

Some scientists believe that global warming could shut down this ocean current system by creating an influx of freshwater from melting ice sheets and glaciers into the subpolar North Atlantic Ocean. Since freshwater is less dense than saline water, a significant intrusion of freshwater would lower the density of the surface waters and thus inhibit the sinking motion that drives large-scale thermohaline circulation. It has also been speculated that, as a consequence of large-scale surface warming, such changes could even trigger colder conditions in regions surrounding the North Atlantic. Experiments with modern climate models suggest that such an event would be unlikely. Instead, a moderate weakening of the thermohaline circulation might occur that would lead to a dampening of surface warming—rather than actual cooling—in the higher latitudes of the North Atlantic Ocean.

Atmospheric Pressure

Atmospheric pressure and wind are both significant controlling factors of Earth's weather and climate. Although these two physical variables may at first glance appear to be quite different, they are in fact closely related. Wind exists because of horizontal and vertical differences (gradients) in pressure, yielding a correspondence that often makes it possible to use the pressure distribution as an alternative representation of atmospheric motions. Pressure is the force exerted on a unit area, and atmospheric pressure is equivalent to the weight of air above a given area on Earth's surface or within its atmosphere. This pressure is usually expressed in millibars (mb; 1 mb equals 1,000 dynes per square cm) or in kilopascals (kPa; 1 kPa equals 10,000 dynes per square cm). Distributions of pressure on a map are depicted by a series of curved lines called isobars, each of which connects points of equal pressure.

At sea level the mean pressure is about 1,000 mb (100 kPa), varying by less than 5 percent from this value at any given location or time. Mean sea-level pressure values for the mid-winter months in the Northern Hemisphere are summarized in this first diagram, and mean sea-level pressure values for the mid-summer months are illustrated in the next diagram. Since charts of atmospheric pressure often represent average values over several days, pressure features that are relatively consistent day after day emerge, while more transient, short-lived features are removed. Those that remain are known as semipermanent pressure centres and are the source regions for major, relatively uniform bodies of air known as air masses. Warm, moist maritime tropical (mT) air forms over tropical and subtropical ocean waters in association with the high-pressure regions prominent there. Cool, moist maritime polar (mP) air, on the other hand, forms over the colder subpolar ocean waters just south and east of the large, winter oceanic low-pressure regions. Over the continents, cold dry continental polar (cP) air and extremely cold dry continental arctic (cA) air forms in the high-pressure regions that are especially pronounced in winter, while hot dry continental tropical (cT) air forms over hot desertlike continental domains in summer in association with low-pressure areas, which are sometimes called heat lows.

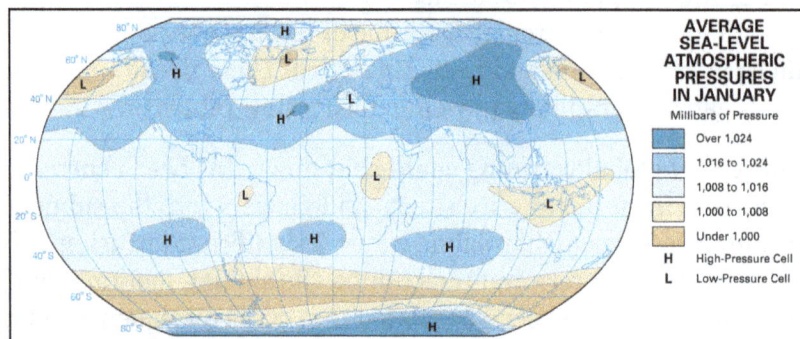

Average atmospheric pressure at sea level.

A closer examination of the diagrams above reveals some interesting features. First, it is clear that sea-level pressure is dominated by closed high- and low-pressure centres, which are largely caused by differential surface heating between low and high latitudes and between continental and oceanic regions. High pressure tends to be amplified over the colder surface features. Second, because of seasonal changes in surface heating, the pressure centres exhibit seasonal changes in their characteristics. For example, the Siberian High, Aleutian Low, and Icelandic Low that are so prominent in the winter virtually disappear in summer as the continental regions warm relative to surrounding bodies of water. At the same time, the Pacific and Atlantic highs amplify and migrate northward.

At altitudes well above Earth's surface, the monthly average pressure distributions show much less tendency to form in closed centres but rather appear as quasi-concentric circles around the poles. This more symmetrical appearance reflects the dominant role of meridional (north-south) differences in radiative heating and cooling. Excess heating in tropical latitudes, in contrast to polar areas, produces higher pressure at upper levels in the tropics as thunderstorms transfer air to higher levels. In addition, the greater heating/cooling contrast in winter yields stronger pressure differences during this season. Perfect symmetry between the tropics and the poles is interrupted by wavelike atmospheric disturbances associated with migratory and semipermanent high- and low-pressure surface weather systems. These weather systems are most pronounced over the Northern Hemisphere, with its more prominent land-ocean contrasts and orographic (high-elevation) features.

Winds

Relationship of Wind to Pressure and Governing Forces

The changing wind patterns are governed by Newton's second law of motion, which states that the sum of the forces acting on a body equals the product of the mass of that body and the acceleration caused by those forces. The basic relationship between atmospheric pressure and horizontal wind is revealed by disregarding friction and any changes in wind direction and speed to yield the mathematical relationship,

$$fu = -\frac{1}{\rho}\frac{\partial p}{\partial y} \text{ and } fv = \frac{1}{\rho}\frac{\partial p}{\partial x},$$

where u is the zonal wind speed (+ eastward), v the meridional wind speed (+ northward), $f = 2\omega \sin \phi$ (Coriolis parameter), ω the angular velocity of Earth's rotation, ϕ the latitude, ρ the air density (mass per unit volume), p the pressure, and x and y the distances toward the east and north, respectively. This simple non-accelerating flow is known as geostrophic balance and yields a motion field known as the geostrophic wind. Equation $fu = -\frac{1}{\rho}\frac{\partial p}{\partial y}$ and $fv = \frac{1}{\rho}\frac{\partial p}{\partial x}$, expresses, for both the x and y directions, a balance between the force created by horizontal differences in pressure (the horizontal pressure-gradient force) and an apparent force that results from Earth's rotation (the Coriolis force). The pressure-gradient force expresses the tendency of pressure differences to effectuate air movement from higher to lower pressure. The Coriolis force arises because the air motions are observed on a rotating nearly spherical body. The total motion of a parcel of air has two parts: $fu = -\frac{1}{\rho}\frac{\partial p}{\partial y}$ and $fv = \frac{1}{\rho}\frac{\partial p}{\partial x}$, the motion relative to Earth as if the planet were fixed, and $p = \rho RT$ the motion given to the parcel of air by the planet's rotation. When the atmosphere is viewed from a fixed point in space, Earth's rotation is apparent. An observer in space would witness the total motion of the atmosphere. Conversely, an observer on the ground sees and measures only the relative motion of the atmosphere, because he is also rotating and cannot see directly the rotational motion applied by Earth. Instead, the observer on the ground sees the effect of the rotation as a deviation applied to the relative motion. The quantity that describes this deviation is the Coriolis force. Because the Coriolis force results from a ground-level frame of reference on a rotating planet, it is not a true force.

More specifically, the observer on the ground experiences the Coriolis force as a deflection of the relative motion to the right in the Northern Hemisphere and to the left in the Southern Hemisphere. Of particular significance in this simple model of wind-pressure relationships is the fact that the geostrophic wind blows in a direction parallel to the isobars, with the low pressure on the observer's left as he looks downwind in the Northern Hemisphere and on his right in the Southern Hemisphere.

Wind speed increases as the distance between isobars decreases (or pressure gradient increases). Curvature (i.e., changes in wind direction) can be added to this model with relative ease in a flow representation known as the gradient wind. The basic wind-pressure relationships, however, remain qualitatively the same. Of greatest importance is the fact that large-scale, observed winds tend to behave much as the geostrophic- or gradient-flow models predict in most of the atmosphere. The most notable exceptions occur in low latitudes, where the Coriolis parameter becomes very small—equation ($fu = -\dfrac{1}{\rho}\dfrac{\partial p}{\partial y}$ and $fv = \dfrac{1}{\rho}\dfrac{\partial p}{\partial x}$,) cannot be used to provide a reliable wind estimate—and in the lowest kilometre of the atmosphere, where friction becomes important. The friction induced by airflow over the underlying surface reduces the wind speed and alters the simple balance of forces such that the wind blows with a component toward lower pressure.

Cyclones and Anticyclones

Cyclones and anticyclones are regions of relatively low and high pressure, respectively. They occur over most of Earth's surface in a variety of sizes ranging from the very large semipermanent examples described to smaller, highly mobile systems.

Common to both cyclones and anticyclones are the characteristic circulation patterns. The geostrophic-wind and gradient-wind models dictate that, in the Northern Hemisphere, flow around a cyclone—cyclonic circulation—is counterclockwise, and flow around an anticyclone—anticyclonic circulation—is clockwise. Circulation directions are reversed in the Southern Hemisphere. In the presence of friction, the superimposed component of motion toward lower pressure produces a "spiraling" effect toward the low-pressure centre and away from the high-pressure centre.

The cyclones that form outside the equatorial belt, known as extratropical cyclones, may be regarded as large eddies in the broad air currents that flow in the general direction from west to east around the middle and higher latitudes of both hemispheres. They are an essential part of the mechanism by which the excess heat received from the Sun in Earth's equatorial belt is conveyed toward higher latitudes. These higher latitudes radiate more heat to space than they receive from the Sun, and heat must reach them by winds from the lower latitudes if their temperature is to be continually cool rather than cold. If there were no cyclones and anticyclones, the north-south movements of the air would be much more limited, and there would be little opportunity for heat to be carried poleward by winds of subtropical origin. Under such circumstances the temperature of the lower latitudes would increase, and the polar regions would cool; the temperature gradient between them would intensify.

Strong horizontal gradients of temperature are particularly favourable for the formation and development of cyclones. The temperature difference between polar regions and the Equator builds up until it becomes sufficiently intense to generate new cyclones. As their associated cold fronts sweep equatorward and their warm fronts move poleward, the new cyclones reduce the temperature difference. Thus, the wind circulation on Earth represents a balance between the heating effects of solar radiation occurring in the polar regions and at the Equator. Wind circulation, through the effect of cyclones, anticyclones, and other wind systems, also periodically destroys this temperature contrast.

Cyclones of a somewhat different character occur closer to the Equator, generally forming in latitudes between 10° to 30° N and S over the oceans. They generally are known as tropical cyclones when their winds equal or exceed 74 miles (119 km) per hour. They are also known as hurricanes if they occur in the Atlantic Ocean and the Caribbean Sea, as typhoons in the western Pacific Ocean and the China Sea, and as cyclones off the coasts of Australia. These storms are of smaller diameter than the extratropical cyclones, ranging from 100 to 500 km (60 to 300 miles) in diameter, and are accompanied by winds that sometimes reach extreme violence.

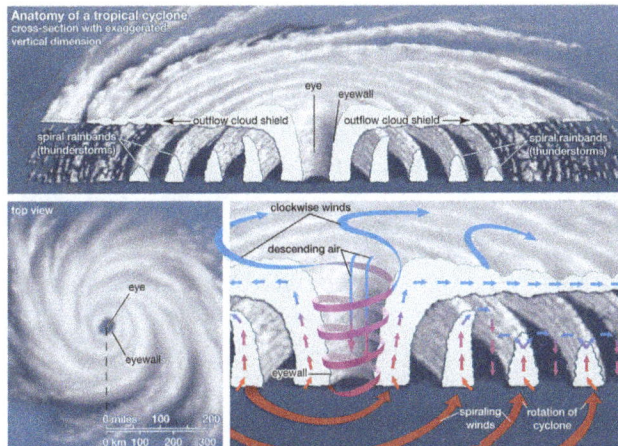

A top view and vertical cross section of a tropical cyclone.

Extratropical Cyclones

Of the two types of large-scale cyclones, extratropical cyclones are the most abundant and exert influence on the broadest scale; they affect the largest percentage of Earth's surface. Furthermore, this class of cyclones is the principal cause of day-to-day weather changes experienced in middle and high latitudes and thus is the focal point of much of modern weather forecasting. The seeds for many current ideas concerning extratropical cyclones were sown between 1912 and 1930 by a group of Scandinavian meteorologists working in Bergen, Nor. This so-called Bergen school, founded by Norwegian meteorologist and physicist Vilhelm Bjerknes, formulated a model for a cyclone that forms as a disturbance along a zone of strong temperature contrast known as a front, which in turn constitutes a boundary between two contrasting air masses. In this model the masses of polar and mid-latitude air around the globe are separated by the polar front (the transition region separating warmer tropical air from colder polar air). This region possesses a strong temperature gradient, and thus it is a reservoir of potential energy that can be readily tapped and converted into the kinetic energy associated with extratropical cyclones.

For this reservoir to be tapped, a cyclone (called a wave, or frontal, cyclone) must develop much in the way shown in the diagram. The feature that is of primary importance prior to cyclone development (cyclogenesis) is a front, represented in the initial stage (A) as a heavy black line with alternating triangles or semicircles attached to it. This stationary or very slow-moving front forms a boundary between cold and warm air and thus is a zone of strong horizontal temperature gradient (sometimes referred to as a baroclinic zone). Cyclone development is initiated as a disturbance along the front, which distorts the front into the wavelike configuration (B; wave appearance). As the pressure within the disturbance continues to decrease, the disturbance assumes the appearance of a cyclone and forces poleward and equatorward movements of warm and cold air, respectively, which are represented by mobile frontal boundaries. As depicted in the cyclonic circulation stage (C), the front that signals the advancing cold air (cold front) is indicated by the triangles, while the front corresponding to the advancing warm air (warm front) is indicated by the semicircles. As the cyclone continues to intensify, the cold dense air streams rapidly equatorward, yielding a cold front with a typical slope of 1 to 50 and a propagation speed that is often 8 to 15 metres per second (about 18 to 34 miles per hour) or more. At the same time, the warm less-dense air moving in a northerly direction flows up over the cold air east of the cyclone to produce a warm front with a typical slope of 1 to 200 and a typically much slower propagation speed of about 2.5 to 8 metres per second (6 to 18 miles per hour). This difference in propagation speeds between the two fronts allows the cold front to overtake the warm front and produce yet another, more complicated frontal structure, known as an occluded front. An occluded front (D) is represented by a line with alternating triangles and semicircles on the same side. This occlusion process may be followed by further storm intensification. The separation of the cyclone from the warm air toward the Equator, however, eventually leads to the storm's decay and dissipation (E) in a process called cyclolysis.

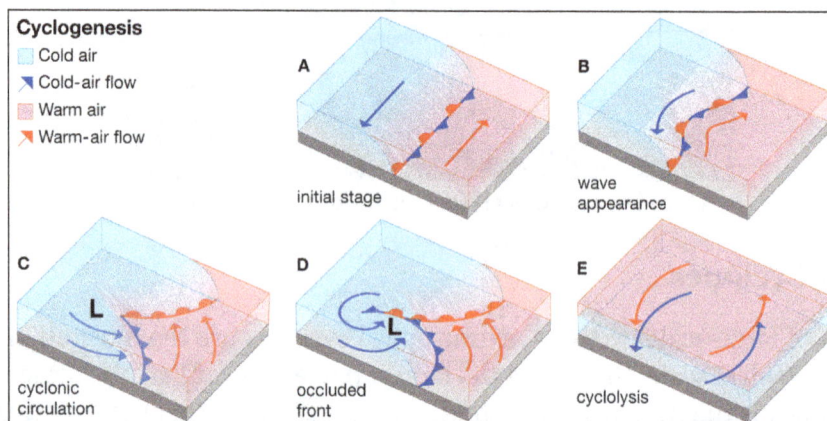

Evolution of a wave (frontal) cyclone.

The life cycle of such an event is typically several days, during which the cyclone may travel from several hundred to a few thousand kilometres. In its path and wake occur dramatic weather changes. A typical sequence of weather possibly resulting from the approach and passage of a cyclone and its fronts through an area is depicted in the diagram. Shown in the occluded-front stage of the cyclogenesis diagram is a cross section of the clouds and precipitation that usually occur along line ab. Warm frontal weather is most frequently characterized by stratiform clouds, which ascend as the front approaches and potentially yield rain or snow. The passing of a warm front brings a rise in air temperature and clearing skies. The warmer air, however, may also harbour the ingredients for rain shower or thunderstorm formation, a condition that is enhanced as the cold front approaches.

The passage of the cold front is marked by the influx of colder air, the formation of stratocumulus clouds with some lingering rain or snow showers, and then eventual clearing. While this is an oft-repeated scenario, it is important to recognize that many other weather sequences can also occur. For example, the stratiform clouds of a warm front may have imbedded cumulus formations and thunderstorms; the warm sector might be quite dry and yield few or no clouds; the pre-cold-front weather may closely resemble that found ahead of the warm front; or the post-cold-front air may be completely cloud-free. Cloud patterns oriented along fronts and spiraling around the cyclone vortex are consistently revealed in satellite pictures of Earth.

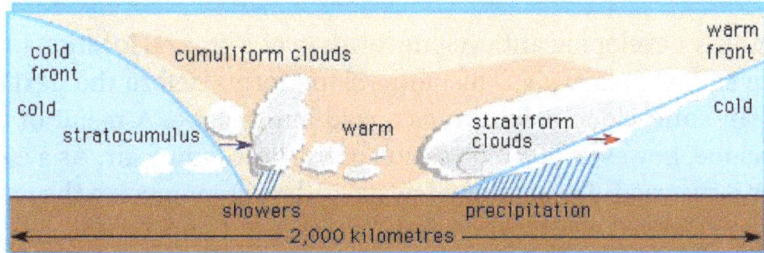

Cross section of clouds and precipitation often found along the cross-sectional line ab in panel D in the cyclogenesis diagram; the direction of frontal movement is indicated by the arrows.

The actual formation of any area of low pressure requires that mass in the column of air lying above Earth's surface be reduced. This loss of mass then reduces the surface pressure. In the late 1930s and early '40s, three members of the Bergen school—Norwegian American meteorologists Jacob Bjerknes and Jørgen Holmboe and Swedish American meteorologist Carl-Gustaf Rossby—recognized that transient surface disturbances were accompanied by complementary wave features in the flow in the middle and higher atmospheric layers associated with the jet stream. These wave features are accompanied by regions of mass divergence and convergence that support the growth of surface-pressure fields and direct their movement.

While extratropical cyclones form and intensify in association with fronts, there are small-scale cyclones that appear in the middle of a single air mass. A notable example is a class of cyclones, generally smaller than the frontal variety, that form in polar air streams in the wake of a frontal cyclone. These so-called polar lows are most prominent in subpolar marine environments and are thought to be caused by the transfer of heat and moisture from the warmer water surface into the overlying polar air and by supporting middle-tropospheric circulation features. Other small-scale cyclones form on the lee side of mountain barriers as the general westerly flow is disturbed by the mountain. These "lee cyclones" may produce major windstorms and dust storms downstream of a mountain barrier.

Satellite image of a large dust storm in the Takla Makan Desert, northwestern China.

Anticyclones

While cyclones are typically regions of inclement weather, anticyclones are usually meteorologically quiet regions. Generally larger than cyclones, anticyclones exhibit persistent downward motions and yield dry stable air that may extend horizontally many hundreds of kilometres.

In most cases, an actively developing anticyclone forms over a ground location in the region of cold air behind a cyclone as it moves away. This anticyclone forms before the next cyclone advances into the area. Such an anticyclone is known as a cold anticyclone. A result of the downward air motion in an anticyclone, however, is compression of the descending air. As a consequence of this compression, the air is warmed. Thus, after a few days, the air composing the anticyclone at levels 2 to 5 km (1 to 3 miles) above the ground tends to increase in temperature, and the anticyclone is transformed into a warm anticyclone.

Warm anticyclones move slowly, and cyclones are diverted around their periphery. During their transformation from cold to warm status, anticyclones usually move out of the main belt followed by cyclones in middle latitudes and often amalgamate with the quasi-permanent bands of relatively high pressure found in both hemispheres around latitude 20° to 30°—the so-called subtropical anticyclones. On some occasions the warm anticyclones remain in the belt normally occupied by the mid-latitude westerly winds. The normal cyclone tracks are then considerably modified; atmospheric depressions (areas of low pressure) are either blocked in their eastward progress or diverted to the north or south of the anticyclone. Anticyclones that interrupt the normal circulation of the westerly wind in this way are called blocking anticyclones, or blocking highs. They frequently persist for a week or more, and the occurrence of a few such blocking anticyclones may dominate the character of a season. Blocking anticyclones are particularly common over Europe, the eastern Atlantic, and the Alaskan area.

The descent and warming of the air in an anticyclone might be expected to lead to the dissolution of clouds and the absence of rain. Near the centre of the anticyclone, the winds are light and the air can become stagnant. Air pollution can build up as a result. The city of Los Angeles, for example, often has poor air quality because it is frequently under a stationary anticyclone. In winter the ground cools, and the lower layers of the atmosphere also become cold. Fog may be formed as the air is cooled to its dew point in the stagnant air. Under other circumstances, the air trapped in the first kilometre above Earth's surface may pick up moisture from the sea or other moist surfaces, and layers of cloud may form in areas near the ground up to a height of about 1 km (0.6 mile). Such layers of cloud can be persistent in anticyclones (except over the continents in summer), but they rarely grow thick enough to produce rain. If precipitation occurs, it is usually drizzle or light snow.

Anticyclones are often regions of clear skies and sunny weather in summer; at other times of the year, cloudy and foggy weather—especially over wet ground, snow cover, and the ocean—may be more typical. Winter anticyclones produce colder than average temperatures at the surface, particularly if the skies remain clear. Anticyclones are responsible for periods of little or no rain, and such periods may be prolonged in association with blocking highs.

Cyclone and Anticyclone Climatology

Migrating cyclones and anticyclones tend to be distributed around certain preferred regions, known as tracks, that emanate from preferred cyclogenetic and anticyclogenetic regions. The contrast between the winter and summer mean sea-level pressure diagrams also indicates the typical cyclone tracks for both January and July. Favoured cyclogenetic regions in the Northern Hemisphere are found on the lee side of mountains and off the east coasts of continents. Cyclones then track east or southeast before eventually turning toward the northeast and decaying. The tracks are displaced farther northward in July, reflecting the more northward position of the polar front in summer. Continental cyclones usually intensify at a rate of 0.5 mb (0.05 kPa) per hour or less, although more dramatic examples can be found. Marine cyclones, on the other hand, often experience explosive development in excess of 1 mb (0.1 kPa) per hour, particularly in winter.

Anticyclones tend to migrate equatorward out of the cold air mass regions and then eastward before decaying or merging with a warm anticyclone. Like cyclones, warm anticyclones also slowly migrate poleward with the warm season.

In the Southern Hemisphere, where most of Earth's surface is covered by oceans, the cyclones are distributed fairly uniformly through the various longitudes. Typically, cyclones form initially in latitudes 30° to 40° S and move in a generally southeastward direction, reaching maturity in latitudes near 60° S. Thus, the Antarctic continent is usually ringed by a number of mature or decaying cyclones. The belt of ocean from 40° to 60° S is a region of persistent, strong westerly winds that form part of the circulation to the north of the main cyclone centres; These are the "roaring forties," where the westerly winds are interrupted only at intervals by the passage southeastward of developing cyclones.

Local Winds

Scale Classes

Organized wind systems occur in spatial dimensions ranging from tens of metres to thousands of kilometres and possess residence times that vary from seconds to weeks. The concept of scale considers the typical size and lifetime of a phenomenon. Since the atmosphere exhibits such a large variety of both spatial and temporal scales, efforts have been made to group various phenomena into scale classes. The class describing the largest and longest-lived of these phenomena is known as the planetary scale. Such phenomena are typically a few thousand kilometres in size and have lifetimes ranging from several days to several weeks. Examples of planetary-scale phenomena include the semipermanent pressure centres and certain globe-encircling upper-air waves.

A second class is known as the synoptic scale. Spanning smaller distances, a few hundred to a few thousand kilometres, and possessing shorter lifetimes, a few to several days, this class contains the migrating cyclones and anticyclones that control day-to-day weather changes. Sometimes the planetary and synoptic scales are combined into a single classification termed the large-scale, or macroscale. Large-scale wind systems are distinguished by the predominance of horizontal motions over vertical motions and by the preeminent importance of the Coriolis force in influencing wind characteristics. Examples of large-scale wind systems include the trade winds and the westerlies.

There is a third class of phenomena of even smaller size and shorter lifetime. In this class, vertical motions may be as significant as horizontal movement, and the Coriolis force often plays a less important role. Known as the mesoscale, this class is characterized by spatial dimensions of ten to a few hundred kilometres and lifetimes of a day or less. Because of the shorter time scale and because the other forces may be much larger, the effect of the Coriolis force in mesoscale phenomena is sometimes neglected.

Two of the best-known examples of mesoscale phenomena are the thunderstorm and its devastating by-product, the tornado. The present discussion focuses on less intense, though nevertheless commonly observed, wind systems that are found in rather specific geographic locations and thus are often referred to as local wind systems.

Local Wind Systems

The so-called sea and land breeze circulation is a local wind system typically encountered along coastlines adjacent to large bodies of water and is induced by differences that occur between the heating or cooling of the water surface and the adjacent land surface. Water has a higher heat capacity (i.e., more units of heat are required to produce a given temperature change in a volume of water) than do the materials in the land surface. Daytime solar radiation penetrates to several metres into the water, the water vertically mixes, and the volume is slowly heated. In contrast, daytime solar radiation heats the land surface more quickly because it does not penetrate more than a few centimetres below the land surface. The land surface, now at a higher temperature relative to the air adjacent to it, transfers more heat to its overlying air mass and creates an area of low pressure. Thus, a circulation cell much like that depicted in the diagram is induced. It should be noted that the surface flow is from the water toward the land and thus is called a sea breeze.

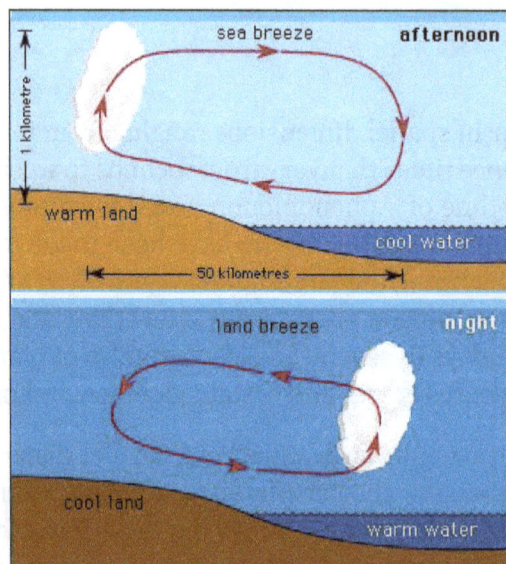

Typical sea-breeze (afternoon) and land-breeze (night) circulations with associated cloud formations.

Since the landmass possesses a lower heat capacity than water, the land cools more rapidly at night than does the water. Consequently, at night the cooler landmass yields a cooler overlying air mass and creates a zone of relatively higher pressure. This produces a circulation cell with air motions opposite to those found during the day. This flow from land to water is known as a land breeze. The land breeze is typically shallower than the sea breeze since the cooling of the atmosphere over land is confined to a shallower layer at night than the heating of the air during the day.

Sea and land breezes occur along the coastal regions of oceans or large lakes in the absence of a strong large-scale wind system during periods of strong daytime heating or nighttime cooling. Those who live within 10 to 20 km (6 to 12 miles) of the coastline often experience the cooler 19- to 37-km-per-hour (12- to 23-mile-per-hour) winds of the sea breeze on a sunny afternoon only to find it turn into a sultry land breeze late at night. One of the features of the sea and land breeze is a region of low-level air convergence in the termination region of the surface flow. Such convergence often induces local upward motions and cloud formations. Thus, in sea and land breeze regions, it is not uncommon to see clouds lying off the coast at night; these clouds are then dissipated by the daytime sea breeze, which forms new clouds, perhaps with showers occurring over land in the afternoon.

Another group of local winds is induced by the presence of mountain and valley features on Earth's surface. One subset of such winds, known as mountain winds or breezes, is induced by differential heating or cooling along mountain slopes. During the day, solar heating of the sun-lit slopes causes the overlying air to move upslope. These winds are also called anabatic flow. At night, as the slopes cool, the direction of airflow is reversed, and cool downslope drainage motion occurs. Such winds may be relatively gentle or may occur in strong gusts, depending on the topographic configuration. These winds are one type of katabatic flow. In an enclosed valley, the cool air that drains into the valley may give rise to a thick fog condition. Fog persists until daytime heating reverses the circulation and creates clouds associated with the upslope motion at the mountain top.

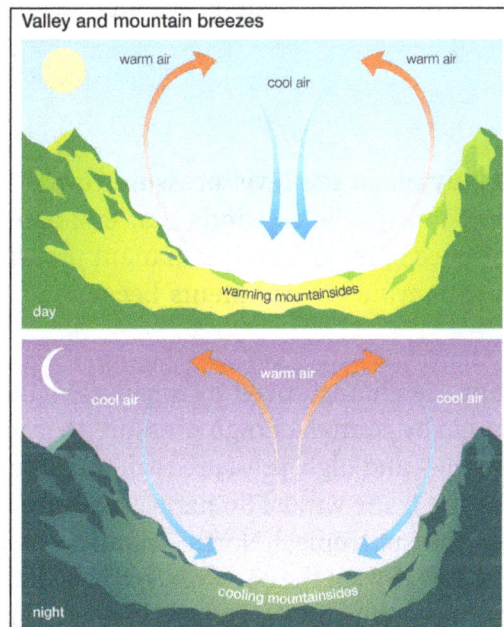

When the valley floor warms during the day, warm air rises up the slopes of surrounding mountains and hills to create a valley breeze. At night, denser cool air slides down the slopes to settle in the valley, producing a mountain breeze.

Another subset of katabatic flow, called foehn winds (also known as chinook winds east of the Rocky Mountains and as Santa Ana winds in southern California), is induced by adiabatic temperature changes occurring as air flows over a mountain. Adiabatic temperature changes are those that occur without the addition or subtraction of heat; they occur in the atmosphere when bundles of air are moved vertically. When air is lifted, it enters a region of lower pressure and expands.

This expansion is accompanied by a reduction of temperature (adiabatic cooling). When air subsides, it contracts and experiences adiabatic warming. As air ascends on the windward side of the mountain, its cooling rate may be moderated by heat that is released during the formation of precipitation. However, having lost much of its moisture, the descending air on the leeward side of the mountain adiabatically warms faster than it was cooled on the windward ascent. Thus, the effect of this wind, if it reaches the surface, is to produce warm, dry conditions. Usually, such winds are gentle and produce a slow warming. On occasion, however, foehn winds may exceed 185 kilometres (115 miles) per hour and produce air-temperature increases of tens of degrees (sometimes more than 20 °C [36 °F]) within only a few hours.

Other types of katabatic wind can occur when the underlying geography is characterized by a cold plateau adjacent to a relatively warm region of lower elevation. Such conditions are satisfied in areas in which major ice sheets or cold elevated land surfaces border warmer large bodies of water. Air over the cold plateau cools and forms a large dome of cold dense air. Unless held back by background wind conditions, this cold air will spill over into the lower elevations with speeds that vary from gentle (a few kilometres per hour) to intense (93 to 185 km [58 to 115 miles] per hour), depending on the incline of the slope of the terrain and the distribution of the background pressure field. Two special varieties of katabatic wind are well known in Europe. One is the bora, which blows from the highlands of Croatia, Bosnia and Herzegovina, and Montenegro to the Adriatic Sea; the other is the mistral, which blows out of central and southern France to the Mediterranean Sea. Creating blizzard conditions, intense katabatic winds often blow northward off the Antarctic Ice Sheet.

Zonal Surface Winds

The diagrams of January and July mean sea-level pressure reveal that, on the average, certain geographic locations can expect to experience winds that emanate from one prevailing direction largely dictated by the presence of major semipermanent pressure systems. Such prevailing winds have long been known in marine environments because of their influence on the great sailing ships.

Tropical and subtropical regions are characterized by a general band of low pressure lying near the Equator. This band is bounded by centres of high pressure that may extend poleward into the middle latitudes. Between these low- and high-pressure regions is the region of the tropical winds. Of these the most extensive are the trade winds. So named because of their favourable influence on trade ships traveling across the subtropical North Atlantic, trade winds flow westward and somewhat in the direction of the Equator on the equatorward side of the subtropical high-pressure centres. The "root of the trades," occurring on the eastern side of a subtropical high-pressure centre, is characterized by subsiding air. This produces the very warm, dry conditions above a shallow layer of oceanic stratus clouds found in the eastern extremes of the subtropical Atlantic and Pacific ocean basins. As the trade winds progress westward, however, subsidence abates, the air mass becomes more humid, and scattered showers appear. These showers occur particularly on islands with elevated terrain features that interrupt the flow of the warm moist air. The equatorward flow of the trade winds of the Northern and Southern hemispheres often results in a convergence of the two air streams in a region known as the intertropical convergence zone (ITCZ). Deep convective clouds, showers, and thunderstorms occur along the ITCZ.

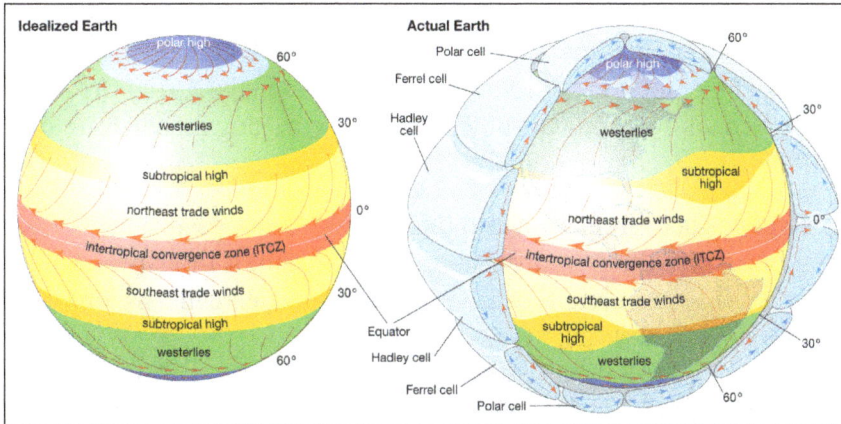

General patterns of atmospheric circulation over an idealized Earth with a uniform surface (left) and the actual Earth (right). Both horizontal and vertical patterns of atmospheric circulation are depicted in the diagram of the actual Earth.

When the air reaches the western extreme of the high-pressure centre, it turns poleward and then eventually returns eastward in the middle latitudes. The poleward-moving air is now warm and laden with moist maritime tropical air (mT); it gives rise to the warm, humid, showery climate characteristic of the Caribbean region, eastern South America, and the western Pacific island chains. The westerlies are associated with the changeable weather common to the middle latitudes. Migrating extratropical cyclones and anticyclones associated with contrasting warm moist air moving poleward from the tropics and cold dry air moving equatorward from polar latitudes yield periods of rain (sometimes with violent thunderstorms), snow, sleet, or freezing rain interrupted by periods of dry, sunny, and sometimes bitterly cold conditions. Furthermore, these patterns are seasonally dependent, with more intense cyclones and colder air prevailing in winter but with a higher incidence of thunderstorms common in spring and summer. In addition, these migrations and the associated climate are complicated by the presence of landmasses and major mountain features, particularly in the Northern Hemisphere.

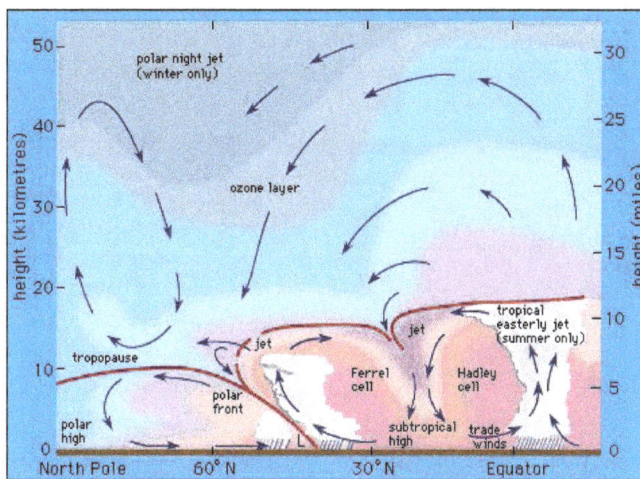

Positions of jet streams in the atmosphere. Arrows indicate directions of mean motions in a meridional plane.

The westerlies lie on the equatorward side of the semipermanent subpolar centres of low pressure. Poleward of these centres, the surface winds turn westward again over significant portions of the subpolar latitudes. As in the middle latitudes, the presence of major landmasses, notably in the

Northern Hemisphere, results in significant variations in these polar easterlies. In addition, the wind systems and the associated climate are seasonally dependent. During the short summer season, the wind systems of the polar latitudes are greatly weakened. During the long winter months, these systems strengthen, and periods of snow alternate with long intervals of dry cold air characteristic of continental polar or continental arctic air masses.

These major regions of surface circulation and their associated pressure fields are related to mean meridional (north-south) circulation patterns as well. Although their presence is discernible in long-term mean statistics accumulated over a hemisphere, such cells are often difficult to detect on a daily basis at any given longitude.

Upper-level Winds

Characteristics

The flow of air around the globe is greatest in the higher altitudes, or upper levels. Upper-level airflow occurs in wavelike currents that may exist for several days before dissipating. Upper-level wind speeds generally occur on the order of tens of metres per second and vary with height. The characteristics of upper-level wind systems vary according to season and latitude and to some extent hemisphere and year. Wind speeds are strongest in the midlatitudes near the tropopause and in the mesosphere.

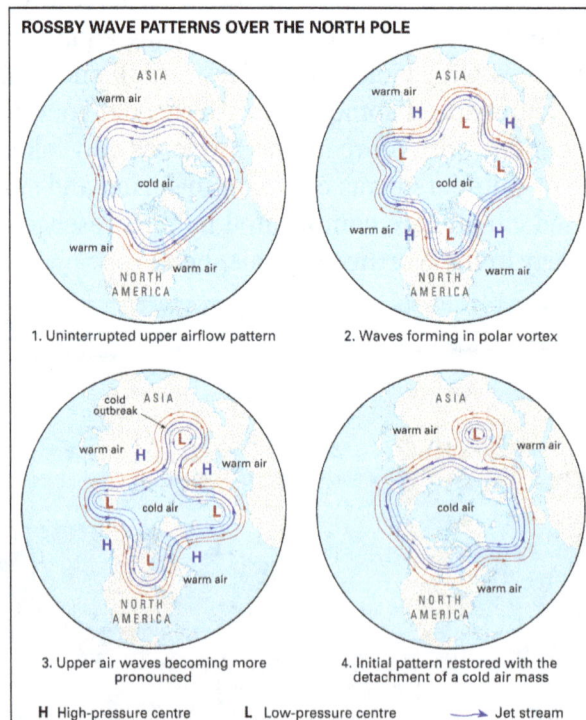

Rossby wave patterns over the North Pole depicting the formation of an outbreak of cold air over Asia.

Upper-level wind systems, like all wind systems, may be thought of as having parts consisting of uniform flow, rotational flow (with cyclonic or anticyclonic curvature), convergent or divergent flow (in which the horizontal area of masses of air shrinks or expands), and deformation (by which the horizontal area of air masses remains constant while experiencing a change in shape).

Upper-level wind systems in the midlatitudes tend to have a strong component of uniform flow from west to east ("westerly" flow), though this flow may change during the summer. A series of cyclonic and anticyclonic vortices superimposed on the uniform west-to-east flow make up a wave train (a succession of waves occurring at periodic intervals). The waves are called Rossby waves after Swedish American meteorologist C.G. Rossby, who first explained fundamental aspects of their behaviour in the 1930s. Waves whose wavelengths are about 6,000 km (3,700 miles) or less are called short waves, while those with longer wavelengths are called long waves. In addition, short waves progress in the same direction as the mean airflow, which is from west to east in the midlatitudes; long waves retrogress (that is, move in the opposite direction of the mean flow). Although the undulating current of air is composed of a number of waves of varying wavelength, the dominant wavelength is usually around several thousand kilometres. Near and underneath the tropopause, regions of divergence are found over regions of gently rising air at the surface, while regions of convergence aloft are found over regions of sinking air below. These regions are usually much more difficult to detect than the regions of rotational and uniform flow. While the horizontal wind speed is typically in the range of 10–50 metres per second (about 20–110 miles per hour), the vertical wind speed associated with the waves is only on the order of centimetres per second.

The characteristics of upper-level wind systems are known mainly from an operational worldwide network of rawinsonde observations. (A rawinsonde is a type of radiosonde designed to track upper-level winds and whose position can be tracked by radar.) Winds measured from Doppler-radar wind profilers, aircraft navigational systems, and sequences of satellite-observed cloud imagery have also been used to augment data from the rawinsonde network; the latter two have been especially useful for defining the wind field over data-sparse regions, such as over the oceans.

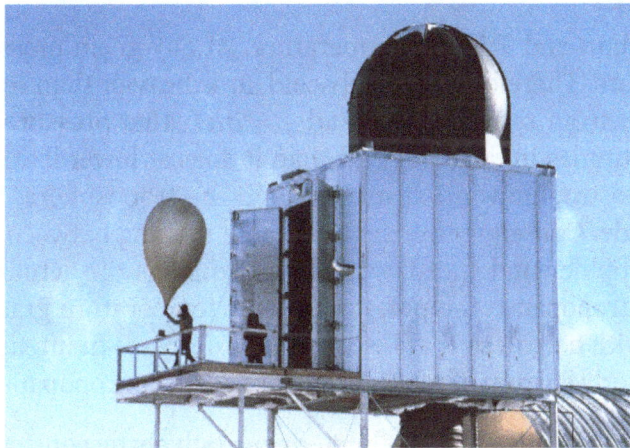

A weather balloon being released at a weather station at the South Pole. Balloon-borne instrument packages designed to track upper-level winds and capable of being tracked by radar are called rawinsondes.

The winds at upper levels, where surface friction does not occur, tend to be approximately geostrophic. In other words, there is a near balance between the pressure gradient force, which directs air from areas of relatively high pressure to areas of relatively low pressure, and the Coriolis force, which deflects air from its straight-line path to the right in the Northern Hemisphere and to the left in the Southern Hemisphere. An important consequence of this geostrophic balance is that the winds blow parallel to isobars (cartographic lines indicating areas of equal pressure), and, according to Buys Ballott's law, lower pressures will be found to the left of the direction of the wind in the Northern Hemisphere and to the right of the wind in the Southern Hemisphere. Furthermore, wind speed increases as the spacing between isobars decreases. In a wave train of westerly flow, the regions of cyclonic flow are associated with troughs of low pressure, whereas anticyclonic flow are characterized by ridges of high pressure. Rising motions tend to be found downstream from the troughs and upstream from the ridges, while sinking motions tend to be found downstream from the ridges and upstream from the troughs. The areas of rising motion tend to be associated with clouds and precipitation (inclement weather), whereas the areas of sinking motion tend to be associated with clear skies (fair weather).

The vertical variation of the structure of the waves depends upon the temperature pattern. In general, because of the net difference in incoming shorter-wavelength solar radiation and outgoing longer-wavelength infrared radiation between the polar and the equatorial regions, there is a horizontal temperature gradient in the troposphere. At both the surface and upper levels, the troposphere is warmest at low latitudes and coldest at high latitudes. The atmosphere is mainly in hydrostatic balance, or equilibrium, between the upward-directed pressure gradient force and the downward-directed force of gravity. This circumstance is expressed in the following relationship:

$$\partial p / \partial z = -\rho g$$

where $\partial p / \partial z$ is the partial derivative of p with respect to z, p is the pressure, z is the height, ρ is the density of the air, and g is the acceleration of gravity. A consequence of this hydrostatic relationship is that the pressure at any level is equal to the weight of the column of air above. According to the ideal gas law,

$$p = \rho R T$$

where R is the gas constant and T is the temperature. At any given pressure, the density varies inversely with temperature. Therefore, relatively cold air is heavier than relatively warm air at the same pressure. It follows from $\partial p / \partial z = -\rho g$ and $p = \rho R T$ that pressure decreases more rapidly with height at high latitudes in the colder air than it does at lower latitudes in the warmer air. If there is a westerly geostrophic wind at midlevels in the troposphere, then pressure decreases with increasing latitude. Consequently, the horizontal spacing between isobars decreases with height. Thus, the geostrophic wind speed, which approximates the actual wind speed, increases with height. Above the tropopause the pole-to-Equator temperature gradient is reversed as air temperature increases with height, so that the westerlies decrease in intensity in the stratosphere. Thus, the strongest westerly current of winds is located near the tropopause.

The aforementioned relationship can be analyzed quantitatively by considering the vertical variation in the geostrophic wind, which is found from the hydrostatic equation $\partial p / \partial z = -\rho g$, the ideal gas law $p = \rho R T$, and the geostrophic wind formula, approximately as follows:

$$\partial u_g / \partial z = -g / fT \, \partial T / \partial y \text{ and } \partial v_g / \partial z = -g / fT \, \partial T / \partial z,$$

where $\partial u_g / \partial z$ is the partial derivative of u_g with respect to z, u_g and v_g are the components of the geostrophic wind in the zonal (straight from west to east) and meridional (north to south) directions, respectively, and f is the Coriolis parameter. The equations given in $\partial u_g / \partial z = -g / fT \, \partial T / \partial y$ and $\partial v_g / \partial z = -g / fT \, \partial T / \partial z$, are known as the thermal wind relations. The difference between the geostrophic wind at some higher level and the geostrophic wind below is called the thermal wind. It follows that the thermal wind vector is oriented so that in the colder air it lies to the left in the Northern Hemisphere and to the right in the Southern Hemisphere.

In addition to the general pole-to-Equator temperature gradient found in the troposphere, there are zonally oriented temperature variations that are wavelike. In fact, to a first approximation, the isotherms (cartographic lines indicating areas of equal temperature) are nearly parallel to the isobars in the upper levels of the troposphere. Most frequently, relatively cold air lies just upstream from upper-level troughs and just downstream from upper-level ridges, while relatively warm air lies just upstream from upper-level ridges and just downstream from upper-level troughs. The thermal wind relation $\partial u_g / \partial z = -g / fT \, \partial T / \partial y$ and $\partial v_g / \partial z = -g / fT \, \partial T / \partial z$, indicates that the wave train of troughs and ridges tilts with height to the west. In the midlatitudes during the summer and in some locations within the midlatitudes during the winter, the meridional temperature gradient weakens so much that the westerlies become weak or nonexistent. As a result, the wavelike wind field disappears and the flow pattern is that of cyclones and anticyclones "cut off" from the flow. When cold air is colocated with the upper-level cyclones and warm air is colocated with the upper-level anticyclones, according to $\partial u_g / \partial z = -g / fT \, \partial T / \partial y$ and $\partial v_g / \partial z = -g / fT \, \partial T / \partial z$, both circulation patterns increase in intensity with height and are called cold-core and warm-core systems, respectively. Tropical cyclones, on the other hand, are warm-core systems that are most intense at the surface and that decrease in intensity with height.

The vertical structure of upper-level waves has an important effect on smaller-scale features that may be embedded within them. The susceptibility of the atmosphere to vertical overturning (a mixing of lower-level warmer air with higher-level colder air) through deep cumulus convection (e.g., thunderstorms) depends on the rate at which temperature decreases with height. When regions of relatively cold air aloft associated with upper-level troughs or cyclones become superimposed during the winter over relatively warm ocean surfaces or during the summer over hot and humid landmasses, then convective storms can form. The type of mesoscale convective system (MCS) that can form depends in large part on the vertical wind shear. When the vertical shear is very strong, supercells and tornadoes may be spawned, especially during the warmer months. During the winter, bands of precipitation sometimes line up along the vertical shear vector through a process known as slantwise convection.

Propagation and Development of Waves

Upper-level waves in the westerlies in midlatitudes usually move from west to east, in part as a result of advection (a process in which the airflow transports a property of the atmosphere [warmth, cold, etc.] downstream) and in part as a result of propagation, which acts in the opposite direction, toward the west. Rossby showed that to a good approximation,

$$c = U - \beta / (2\pi / L)^2,$$

where c is the phase speed of the waves, U is the speed from west to east of the component of upper-level wind due to uniform flow, β is the meridional, or north-south, gradient of the Coriolis parameter (f), and L is the zonal wavelength (the distance between successive troughs or ridges). According to $c = U - \beta / (2\pi / L)^2$, since the magnitude of f increases toward the poles, β is positive, and hence waves whose wavelengths are short have a relatively small component due to propagation. In this situation, advection overwhelms the effect of propagation, and the waves move on downstream. On the other hand, if in midlatitudes the wavelength is very long, then the effects of propagation may exactly cancel the effects of advection, and the waves may become stationary; or if the wavelength becomes even longer, then the waves may become retrograde. From $c=U - \beta / (2\pi / L)^2$, it can be seen that Rossby waves owe their propagation characteristics to the north-south variation of f. In nature, temperature effects and heating and cooling over warm and cold surfaces can modify $c=U - \beta / (2\pi / L)^2$, somewhat.

The physical basis for $c=U - \beta / (2\pi / L)^2$, and for the development of upper-level systems and how they relate to surface systems is described by an elegant theory developed in the late 1940s called quasigeostrophic theory. A measure of the tendency for a fluid to rotate is known as vorticity and is given by the following equation:

$$\zeta = \partial v / \partial x - \partial u / \partial y$$

where ζ is the relative vorticity with respect to Earth's surface. The variables x and y are the coordinate axes for space and correspond to the measurements to the east and north, respectively. The variables u and v are zonal and meridional components (the components of motion in the easterly and northerly directions), respectively, of the wind. On the rotating Earth, the vorticity is the sum of the relative vorticity with respect to Earth's surface, given by the aforementioned expression, and Earth's vorticity, given by f, the Coriolis parameter. Troughs are associated with cyclonic vorticity, and ridges are associated with anticyclonic vorticity. In a wave train, the pressure falls downstream from troughs, where the wind is directed from the region of maximum vorticity along the trough to the region of minimum vorticity, which is along the ridge, and the pressure rises downstream from ridges. On the other hand, pressure can rise east of troughs (and west of ridges) where there is a component of motion from the Equator to the pole. For example, pressure rises from regions of low magnitude off to higher magnitude of f (from low values of Earth's vorticity to higher values of Earth's vorticity). Likewise, pressure falls west of troughs (and east of ridges) where there is a component of motion from one of the poles to the Equator—from relatively high magnitude off to lower magnitude of f. The effect of pressure increases and decreases is greatest when the wavelength is relatively short, such as when the effects of the advection of Earth's vorticity are overwhelmed by the effects of advection of relative vorticity.

The development and amplification of Rossby waves is typically a result of the advection of warmer or colder air at low levels. When warm air is advected underneath a layer of air not experiencing much, if any, advection, the pressure at the top of the layer rises. Conversely, pressure falls when cold air advects under a similar layer of air. If the wave train tilts to the west with height so that cold air lies to the west of troughs and thus east of ridges, the pressure aloft in the troughs decreases. Similarly, when warm air lies to the east of troughs and thus west of ridges, the pressure aloft in the ridges increases. As a result, the amplitude of the waves in the wave train increases, thereby enhancing the temperature advection process, so that there is a positive feedback mechanism that makes the waves continue to amplify. In this process, called baroclinic instability, potential energy is converted into kinetic energy—which occurs as wind—as warm, light air rises and cold, heavy air sinks. Since baroclinic instability is associated with horizontal temperature gradients, according to the thermal wind relation $\partial u_g / \partial z = -g / fT \, \partial T / \partial y$ and $\partial v_g / \partial z = -g / fT \, \partial T / \partial z$, , there must be vertical wind shear.

It is also possible for Rossby waves to amplify through a process called barotropic instability. Barotropic instability, however, requires horizontal shear, not vertical shear; kinetic energy for the waves comes from the mean kinetic energy associated with the westerly wind current. The waves grow in amplitude at the expense of the mean flow. Barotropic instability can occur when the horizontal shear varies with latitude such that the sum of Earth's vorticity and the relative vorticity associated with the horizontal shear is small with respect to latitude.

Relationships to Surface Features

Rossby waves propagating through the upper and middle troposphere cause disturbances to form at the surface. According to quasigeostrophic theory, when there is a wave train embedded within a zone of pole-to-Equator temperature gradient, air rises east of upper-level troughs (and west of upper-level ridges) and sinks west of upper-level troughs (and east of upper-level ridges). These vertical air motions are required to maintain the approximate geostrophic and hydrostatic balance, which are necessary for quasigeostrophic equilibrium. Air converges at the surface underneath the rising current of air to compensate for the upward loss of mass and diverges at the surface underneath a sinking current of air to compensate for the downward gain of mass. As a consequence of the lateral deviation of the air by the Coriolis force, Earth's vorticity is converted into cyclonic relative vorticity where air converges and anticyclonic relative vorticity where air diverges. According to the geostrophic wind relation, cyclonic gyres are associated with low-pressure centres, whereas anticyclonic gyres are connected with areas of high pressure. Thus, low-pressure areas form at the surface downstream from upper-level troughs and upstream from upper-level ridges, whereas the reverse is true for high-pressure areas. These surface low- and high-pressure areas thereby create a westward tilt with height of the waves in pressure. Since there tends to be a pole-to-Equator-directed geostrophic wind west of surface lows and east of surface highs, and an Equator-to-pole-directed geostrophic wind east of surface lows and west of surface highs, there is cold advection underneath upper-level troughs and warm advection underneath upper-level ridges; the baroclinic instability process is thus facilitated.

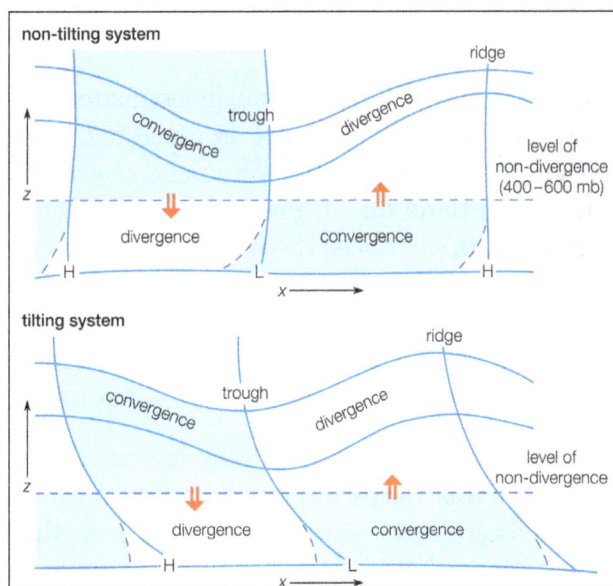

Vertical cross sections through a wave system depicting typical divergence and convergence distributions for non-tilting and tilting systems.

Jet Streams

The upper-level wind flow is frequently concentrated into relatively narrow bands called jet streams, or jets. The jets, whose wind speeds are usually in excess of 30 metres per second (about 70 miles per hour) but can be as high as 107 metres per second (about 240 miles per hour), act to steer upper-level waves. Jet streams are of great importance to air travel because they affect the ground speed, the velocity relative to the ground, of aircraft. Since strong upper-level flow is usually associated with strong vertical wind shear, jet streams in midlatitudes are accompanied by strong horizontal temperature gradients, as required by the thermal wind relation $\partial u_g / \partial z = -g / fT \, \partial T / \partial y$ and $\partial v_g / \partial z = -g / fT \, \partial T / \partial z$. Some regions of high vertical wind shear are marked by clear-air turbulence (CAT). Jet streams whose extents are relatively isolated are called jet streaks. Well-defined circulation patterns of rising and sinking air are usually found just upstream and downstream, respectively, from jet streaks (that are not too curved). Rising motion is found to the left and right just downstream and upstream, respectively, and sinking motion is found to the right and left just downstream and upstream, respectively. Jets tend to be strongest near the tropopause where the horizontal temperature gradient reverses.

The polar front jet moves in a generally westerly direction in midlatitudes, and its vertical wind shear which extends below its core is associated with horizontal temperature gradients that extend to the surface. As a consequence, this jet manifests itself as a front that marks the division between colder air over a deep layer and warmer air over a deep layer. The polar front jet can be baroclinically unstable and break up into waves. The subtropical jet is found at lower latitudes and at slightly higher elevation, because of the increase in height of the tropopause at lower latitudes. The associated horizontal temperature gradients of the subtropical jet do not extend to the surface, so that a surface front is not evident. In the tropics an easterly jet is sometimes found at upper levels, especially when a landmass is located poleward of an ocean, so the temperature increases with latitude. The polar front jet and the subtropical jet play a role in maintaining Earth's general circulation. They are slightly different in each hemisphere because of differences in the distribution of landmasses and oceans.

Winds in the Stratosphere and Mesosphere

The winds in the stratosphere and mesosphere are usually estimated from temperature data collected by satellites. The winds at these high levels are assumed to be geostrophic. Overall, in the midlatitudes, they have a westerly component in the winter and an easterly component in the summer. The highest zonal winds are around 60–70 metres per second (135–155 miles per hour) at 65–70 km (40–43 miles) above Earth's surface. The west-wind component is stronger during the winter in the Southern Hemisphere. The axes of the strongest easterly and westerly wind components in the Southern Hemisphere tilt toward the south with increased altitude during the Northern Hemisphere winter and the Southern Hemisphere summer. The zonal component of the thermal wind shear is in accord with the zonal distribution of temperature.

During the winter there is, in the mean, an intense cyclonic vortex about the poles in the lower stratosphere. Over the North Pole this vortex has an embedded mean trough over northeastern North America and over northeastern Asia, whereas over the Pacific there is a weak anticyclonic vortex. The winter cyclonic vortex over the South Pole is much more symmetrical than the one over the North Pole. During the summer there is an anticyclone above each pole that is much weaker than the wintertime cyclone.

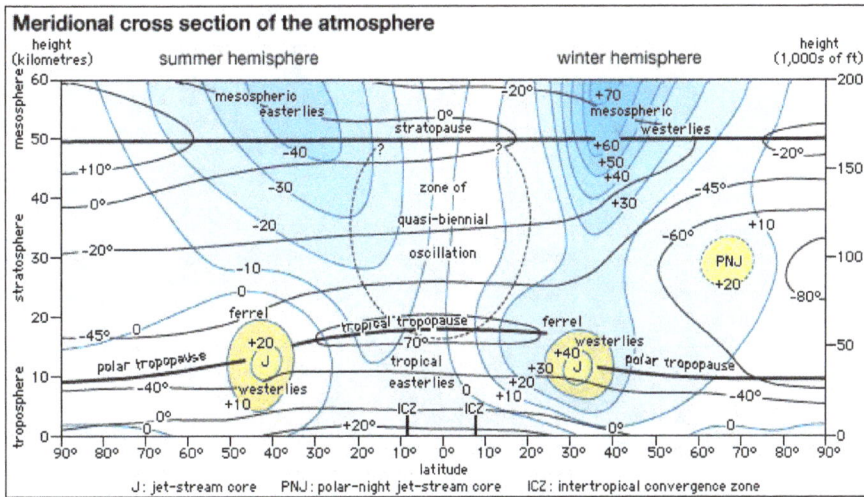

In the figure, Meridional cross section of the atmosphere to a height of 60 km (37 miles) in Earth's summer and winter hemispheres, showing seasonal changes. Numerical values for wind are in units of metres per second and are typical of the Northern Hemisphere, but the structure is much the same in the Southern Hemisphere. Positive and negative signs indicate winds of opposite direction.

In the stratosphere, deviations from the mean behaviour of the winds occur during events called sudden warmings, when the meridional temperature gradient reverses on timescales as short as several days. This also has the effect of reversing the zonal wind direction. Sudden warmings tend to occur during the early and middle parts of the winter and the transition period from winter to spring. The latter marks the changeover from the cold winter polar cyclone to the warm summer polar anticyclone. It is noteworthy that long waves from the troposphere can propagate into the stratosphere during the winter when westerlies and sudden warmings occur, but this is not the case during the summer when easterly winds prevail.

The zonal component of the winds in the stratosphere above equatorial and tropical regions is, in the mean, relatively weak. This is not necessarily the case at any given time, because they reverse direction on the average every 13–14 months. This phenomenon, which is known as the quasi-biennial oscillation (QBO), is caused by the interaction of vertically propagating waves with the mean flow. Its effect is greatest about 27 km (17 miles) above Earth's surface in the equatorial region. The strongest easterlies are stronger than the strongest westerlies.

Monsoons as Strong Seasonal Pressure Variations

Particularly strong seasonal pressure variations occur over continents, as shown in the January and July maps of sea-level atmospheric pressure. Such seasonal fluctuations, commonly called monsoons, are more pronounced over land surfaces because these surfaces are subject to more significant seasonal temperature variations than are water bodies. Since land surfaces both warm and cool faster than water bodies, they often quickly modify the temperature and density characteristics of air parcels passing over them.

Monsoon storm.

Monsoons blow for approximately six months from the northeast and six months from the south-west, principally in South Asia and parts of Africa; however, similar conditions also occur in Central America and the area between Southeast Asia and Australia. Summer monsoons have a dominant westerly component and a strong tendency to converge, rise, and produce rain. Winter monsoons have a dominant easterly component and a strong tendency to diverge, subside, and cause drought. Both are the result of differences in annual temperature trends over land and sea.

Diurnal Variability

Landmasses in regions affected by monsoons warm up very rapidly in the afternoon hours, especially on days with cloud-free conditions; surface air temperatures between 35 and 40 °C (95 and 104 °F) are not uncommon. Under such conditions, warm air is slowly and continually steeped in the moist and cloudy environment of the monsoon. Consequently, over the course of a 24-hour period, energy from this pronounced diurnal, or daily, change in terrestrial heating is transferred to the cloud, rain, and diurnal circulation systems. The scale of this diurnal change extends from that of coastal sea breezes to that of continent-sized processes. Satellite observations have confirmed that the effects of rapid diurnal temperature change occur at continental scales. For example, air from surrounding areas is drawn into the lower troposphere over warmer land areas of South Asia during summer afternoon hours. This buildup of afternoon heating is accompanied by the production of clouds and rain. In contrast, a reverse circulation, characterized by suppressed clouds and rain, is noted in the early morning hours.

Intra-annual Variability

Monsoon rainfall and dry spells alternate on several timescales. One such well-known timescale is found around periods of 40–50 or 30–60 days. This is called the Madden-Julian oscillation (MJO), named for American atmospheric scientists Roland Madden and Paul Julian in 1971. This phenomenon comes in the form of alternating cyclonic and anticyclonic regions that enhance and suppress rainfall, respectively, and flow eastward along the Equator in the Indian and Pacific oceans. The MJO has the ability to influence monsoonal circulation and rainfall by adding moisture during its cyclonic (wet) phase and reducing convection during its anticyclonic (dry) phase. At the surface in monsoon regions, both dry and wet spells result. These periods may alternate locally on the order of two or more weeks per phase.

Interannual Variability

The variability of monsoon-driven rainfall in the Indian Ocean and Australia appears to parallel El Niño episodes. During El Niño events, which occur about every two to seven years, ocean temperatures rise over the central equatorial Pacific Ocean by about 3 °C (5.4 °F). Atypical conditions characterized by increased rising air motion, convection, and rain are created in the western equatorial Pacific. At the same time, a compensating lobe of descending air, producing below-normal rainfall, appears in the vicinity of eastern Australia, Malaysia, and India. The graph illustrates a well-known El Niño–monsoon rainfall relationship. Here, precipitation figures from above- and below-normal monsoon rainfall periods over India are expressed as a function of years. Years characterized by El Niño events are marked by darkened histogram barbs. The graph shows that many of the years with below-normal monsoon rainfall coincide with El Niño years. This illustration provides only limited guidance to seasonal forecasters since monsoon rainfall is close to normal during many El Niño and La Niña years.

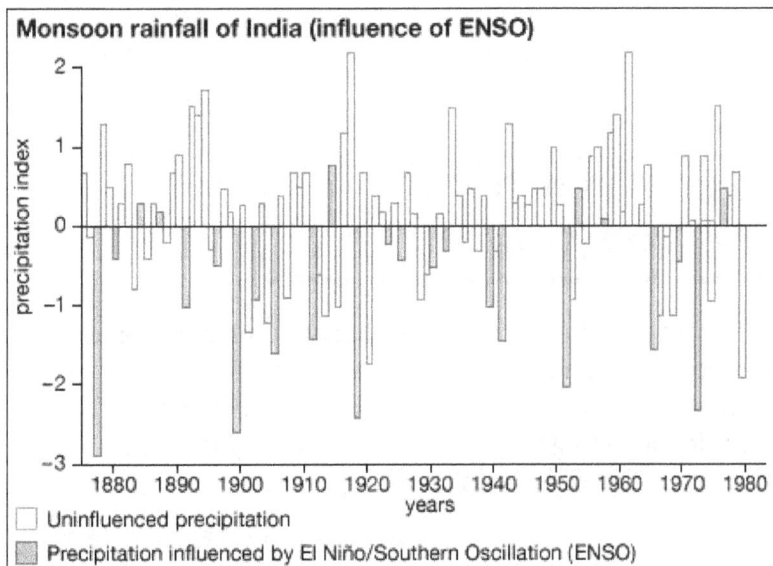

Graph depicting the influence of El Niño/Southern Oscillation (ENSO) on rainfall produced by the Indian summer monsoon. During years when ENSO is active, monsoon-driven precipitation over India often declines.

Many other factors, aside from equatorial Pacific Ocean surface temperatures, contribute to the interannual variability of monsoon rainfall. Excessive spring snow and ice cover on the Plateau of Tibet is related to the deficient monsoon rainfall that occurs during the following summer season in India. Furthermore, strong evidence exists that relates excessive snow and ice cover in western Siberia to deficient Asian summer rainfall. Warmer than normal sea surface temperatures over the Indian Ocean may also contribute somewhat to above-normal rainfall in South Asia. The interplay among these many factors makes forecasting monsoon strength a challenging problem for researchers.

A rather clear signature on the decadal variability of Indian rainfall has been documented by the Indian Weather Service. Decadal-scale variability appears in the graph as an annual running mean that combines average rainfall anomalies (totals as a departure from normal rainfall amounts) occurring at all Indian rain gauge sites. Periods of heavier-than-normal rainfall are followed by decades of somewhat less rainfall.

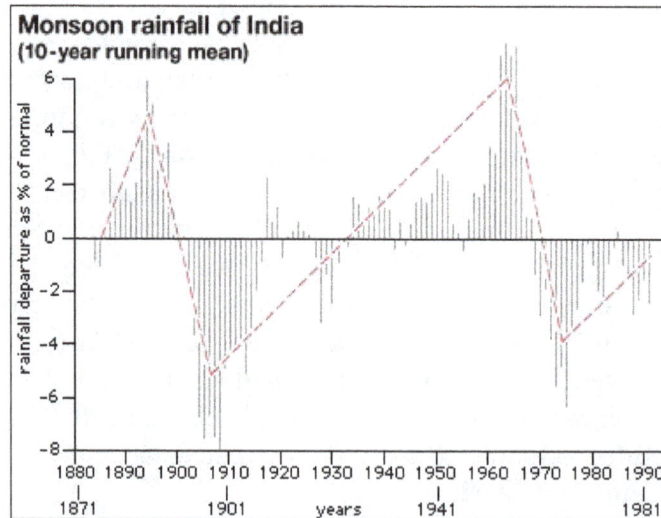

Graph of monsoon rainfall in India, 1871–1981. Annual rainfall amounts are depicted as percentages departing from the 110-year average. The red line superimposed on the graph suggests a recurring trend over this time period.

Affect of Large Water Bodies on Climate of Coastal Areas

Large bodies of water change temperature slower than land masses. Land masses near large bodies of water, especially oceans, change temperature as the oceans change temperature: slower and with less extreme fluctuations than land masses farther away. Ocean currents like the Gulf Stream carry heat from the tropics, affecting the climate of areas away from the tropics. Warm water also increases evaporation and ultimately precipitation.

Water Stores Energy

Water has a much greater ability to store heat than many other substances. On average, the amount of energy needed to increase the temperature of a body of water by 1 degree Celsius (a span of 1.8 degrees Fahrenheit) is about 4 1/2 times greater than the amount needed to heat up an equal mass of land. Consequently, large bodies of water heat up and cool down more slowly than adjacent land masses, so their temperature changes less dramatically with the seasons.

Seasonal Changes

In regions north or south of the tropics, large bodies of water like the ocean release heat during the winter and soak it up during the summer, keeping temperatures within a more moderate range. In other words, the ocean acts a little like a heat sink - and a very effective one at that. The uppermost 10 feet of the ocean can store as much heat as Earth's entire atmosphere.

Ocean Currents

The oceans play a complicated role in coastal climates due to ocean currents, which act as giant conveyor belts transporting heat away from tropical regions toward the frigid poles. Often this serves to keep coastal regions at northern latitudes warmer than they would be otherwise. The famous Gulf Stream, for example, transports heat north along the eastern coast of North America and ultimately toward Europe, ensuring that Europe enjoys a warmer and more moderate climate than it would without the stream.

Tropical Regions

In tropical regions, both land and oceans remain warm year-round. The warm ocean waters give rise to tropical storms called cyclones or hurricanes, a feature of the tropics that can have a devastating effect on coastal regions. As masses of water vapor rise from warm ocean water, the air becomes saturated and the water begins to condense, releasing huge amounts of heat so that the ocean surface remains warm, driving further evaporation and creating a deadly cycle. This cycle only ends when the hurricane passes over land or cold water, at which point no further moisture is available to fuel its growth.

Precipitation

Precipitation is all liquid and solid water particles that fall from clouds and reach the ground. These particles include drizzle, rain, snow, snow pellets, ice crystals, and hail.

A rain shaft piercing a tropical sunset as seen from Man-o'-War Bay, Tobago, Caribbean Sea.

The essential difference between a precipitation particle and a cloud particle is one of size. An average raindrop has a mass equivalent to about one million cloud droplets. Because of their large size, precipitation particles have significant falling speeds and are able to survive the fall from the cloud to the ground.

The transition from a cloud containing only cloud droplets to one containing a mixture of cloud droplets and precipitation particles involves two basically different steps: the formation of incipient precipitation elements directly from the vapour state and the subsequent growth of those elements through aggregation and collision with cloud droplets. The initial precipitation elements may be either ice crystals or chemical-solution droplets.

Development of precipitation through the growth of ice crystals depends on the fact that cloud droplets can freeze spontaneously at temperatures below about −40 °C, or −40 °F. (The reduction of cloud droplets to temperatures below the normal freezing point is termed supercooling.) Within supercooled clouds, ice crystals may form through sublimation of water vapour on certain atmospheric dust particles known as sublimation nuclei. In natural clouds, ice crystals form at temperatures colder than about −15 °C (+5 °F). The exact temperature of ice crystal formation depends largely on the physical-chemical nature of the sublimation nucleus.

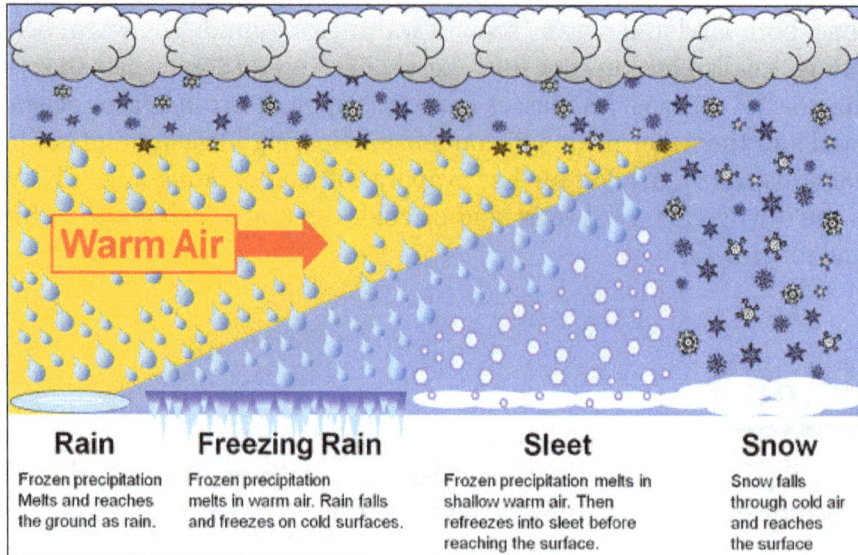

Rain — Frozen precipitation Melts and reaches the ground as rain.

Freezing Rain — Frozen precipitation melts in warm air. Rain falls and freezes on cold surfaces.

Sleet — Frozen precipitation melts in shallow warm air. Then refreezes into sleet before reaching the surface.

Snow — Snow falls through cold air and reaches the surface

Classification of frozen precipitation.

Once ice crystals have formed within a supercooled cloud, they continue to grow as long as their temperature is colder than freezing. The rates of growth depend primarily upon the temperature and degree of vapour saturation of the ambient air. The crystals grow at the expense of the water droplets. In favourable conditions—e.g., in a large, rapidly growing cumulus cloud—an ice crystal will grow to a size of about 0.13 millimetre (0.005 inch) in three to five minutes after formation. At this size, the rate of growth through sublimation slows down, and further growth is largely through aggregation and collision with cloud droplets.

Small solution drops are also important as incipient precipitation particles. The atmosphere contains many small particles of soluble chemical substances. The two most common are sodium chloride swept up from the oceans and sulfate-bearing compounds formed through gaseous reactions in the atmosphere. Such particles, called condensation nuclei, collect water because of their hygroscopic nature and, at relative humidities above about 80 percent, exist as solution droplets. In tropical maritime air masses, the number of condensation nuclei is often very large. Clouds forming in such air may develop a number of large solution droplets long before the tops of the clouds reach temperatures favourable to the formation of ice crystals.

Regardless of whether the initial precipitation particle is an ice crystal or a droplet formed on a condensation nucleus, the bulk of the growth of the precipitation particle is through the mechanisms of collision and coalescence. Because of their larger size, the incipient precipitation elements fall faster than do cloud droplets. As a result, they collide with the droplets lying in their fall path. The rate of growth of a precipitation particle through collision and coalescence is governed by the relative sizes of the particle and the cloud droplets in the fall path that are actually hit by the precipitation particle and the fraction of these droplets that actually coalesce with the particle after collision.

Temperature

Global Variation of Mean Temperature

Global variations of average surface-air temperatures are largely due to latitude, continentality, ocean currents, and prevailing winds.

The effect of latitude is evident in the large north-south gradients in average temperature that occur at middle and high latitudes in each winter hemisphere. These gradients are due mainly to the rapid decrease of available solar radiation but also in part to the higher surface reflectivity at high latitudes associated with snow and ice and low solar elevations. A broad area of the tropical ocean, by contrast, shows little temperature variation.

Continentality is a measure of the difference between continental and marine climates and is mainly the result of the increased range of temperatures that occurs over land compared with water. This difference is a consequence of the much lower effective heat capacities of land surfaces as well as of their generally reduced evaporation rates. Heating or cooling of a land surface takes place in a thin layer, the depth of which is determined by the ability of the ground to conduct heat. The greatest temperature changes occur for dry, sandy soils, because they are poor conductors with very small effective heat capacities and contain no moisture for evaporation. By far the greatest effective heat capacities are those of water surfaces, owing to both the mixing of water near the surface and the penetration of solar radiation that distributes heating to depths of several metres. In addition, about 90 percent of the radiation budget of the ocean is used for evaporation. Ocean temperatures are thus slow to change.

The effect of continentality may be moderated by proximity to the ocean, depending on the direction and strength of the prevailing winds. Contrast with ocean temperatures at the edges of each continent may be further modified by the presence of a north- or south-flowing ocean current. For most latitudes, however, continentality explains much of the variation in average temperature at a fixed latitude as well as variations in the difference between January and July temperatures.

Diurnal, Seasonal and Extreme Temperatures

The diurnal range of temperature generally increases with distance from the sea and toward those places where solar radiation is strongest—in dry tropical climates and on high mountain plateaus (owing to the reduced thickness of the atmosphere to be traversed by the Sun's rays). The average difference between the day's highest and lowest temperatures is 3 °C (5 °F) in January and 5 °C (9 °F) in July in those parts of the British Isles nearest the Atlantic. The difference is 4.5 °C (8 °F) in January and 6.5 °C (12 °F) in July on the small island of Malta. At Tashkent, Uzbekistan, it is 9 °C (16 °F) in January and 15.5 °C (28 °F) in July, and at Khartoum, Sudan, the corresponding figures are 17 °C (31 °F) and 13.5 °C (24 °F). At Kandahār, Afghanistan, which lies more than 1,000 metres (about 3,300 feet) above sea level, it is 14 °C (25 °F) in January and 20 °C (36 °F) in July. There, the average difference between the day's highest and lowest temperatures exceeds 23 °C (41 °F) in September and October, when there is less cloudiness than in July. Near the ocean at Colombo, Sri L., the figures are 8 °C (14 °F) in January and 4.5 °C (8 °F) in July.

The seasonal variation of temperature and the magnitudes of the differences between the same month in different years and different epochs generally increase toward high latitudes and with distance from the ocean. Extreme temperatures observed in different parts of the world are listed in the table.

World Temperature Extremes			
Highest recorded air temperature			
		Temperature	
Continent or region	Place (with elevation)	degrees C	degrees F
Africa	Kebili, Tunisia (38.1 m or 125 ft)	55	131
Antarctica	Vanda Station 77°32′ S 161°40′ E (15 m or 49 ft)	15	59
Asia	Tirat Zevi, Israel (−220 m or −722 ft)	54	129.2
Australia	Oodnadatta, South Australia (112 m or 367 ft)	50.7	123
Europe	Athens, Greece (236 m or 774 ft)	48	118.4
North America	Death Valley (Greenland Ranch), California, U.S. (−54 m or −177 ft)	56.7	134
South America	Rivadavia, Argentina (668 m or 2,192 ft)	48.9	120
Oceania	Tuguegarao, Luzon, Philippines (62 m or 203 ft)	42.2	108
Lowest recorded air temperature			
		Temperature	
Continent or region	Place (with elevation)	degrees C	degrees F
Africa	Ifrane, Morocco (1,635 m or 5,364 ft)	−23.9	−11
Antarctica	Vostok 77°32′ S 106°40′ E (3,420 m or 11,220 ft)	−89.2	−128.6
Asia	Verkhoyansk, Russia (107 m or 351 ft) Oymyakon, Russia (800 m or 2,624 ft)	−67.8	−90

Australia	Charlotte Pass, New South Wales (1,755 m or 5,758 ft)	−23	−9.4
Europe	Ust-Shchuger, Russia (85 m or 279 ft)	−58.1	−72.6
North America	Snag, Yukon, Canada (646 m or 2,119 ft)	−63	−81.4
South America	Sarmiento, Argentina (268 m or 879 ft)	−32.8	−27

Variation with Height

There are two main levels where the atmosphere is heated—namely, at Earth's surface and at the top of the ozone layer (about 50 km, or 30 miles, up) in the stratosphere. Radiation balance shows a net gain at these levels in most cases. Prevailing temperatures tend to decrease with distance from these heating surfaces (apart from the ionosphere and the outer atmospheric layers, where other processes are at work). The world's average lapse rate of temperature (change with altitude) in the lower atmosphere is 0.6 to 0.7 °C per 100 metres (about 1.1 to 1.3 °F per 300 feet). Lower temperatures prevail with increasing height above sea level for two reasons: (1) because there is a less favourable radiation balance in the free air, and (2) because rising air—whether lifted by convection currents above a relatively warm surface or forced up over mountains—undergoes a reduction of temperature associated with its expansion as the pressure of the overlying atmosphere declines. This is the adiabatic lapse rate of temperature, which equals about 1 °C per 100 metres (about 2 °F per 300 feet) for dry air and 0.5 °C per 100 metres (about 1 °F per 300 feet) for saturated air, in which condensation (with liberation of latent heat) is produced by adiabatic cooling. The difference between these rates of change of temperature (and therefore density) of rising air currents and the state of the surrounding air determines whether the upward currents are accelerated or retarded—i.e., whether the air is unstable, so vertical convection with its characteristically attendant tall cumulus cloud and shower development is encouraged or whether it is stable and convection is damped down.

For these reasons, the air temperatures observed on hills and mountains are generally lower than on low ground, except in the case of extensive plateaus, which present a raised heating surface (and on still, sunny days, when even a mountain peak is able to warm appreciably the air that remains in contact with it).

Circulation, Currents and Ocean-atmosphere Interaction

The circulation of the ocean is a key factor in air temperature distribution. Ocean currents that have a northward or southward component, such as the warm Gulf Stream in the North Atlantic or the cold Peru (Humboldt) Current off South America, effectively exchange heat between low and high latitudes. In tropical latitudes the ocean accounts for a third or more of the poleward heat transport; at latitude 50° N, the ocean's share is about one-seventh. In the particular sectors where the currents are located, their importance is of course much greater than these figures, which represent hemispheric averages.

A good example of the effect of a warm current is that of the Gulf Stream in January, which causes a strong east-west gradient in temperatures across the eastern edge of the North American continent. The relative warmth of the Gulf Stream affects air temperatures all the way across the Atlantic, and prevailing westerlies extend the warming effect deep into northern Europe. As a result, January temperatures of Tromsø, Nor. (69°40′ N), for example, average 24 °C (43 °F) above the mean for that latitude. The Gulf Stream maintains a warming influence in July, but it is not as noticeable because of the effects of continentality.

The ocean, particularly in areas where the surface is warm, also supplies moisture to the atmosphere. This in turn contributes to the heat budget of those areas in which the water vapour is condensed into clouds, liberating latent heat in the process. This set of events occurs frequently in high latitudes and in locations remote from the ocean where the moisture was initially taken up.

The great ocean currents are themselves wind-driven—set in motion by the drag of the winds over vast areas of the sea surface, especially where the tops of waves increase the friction with the air above. At the limits of the warm currents, particularly where they abut directly upon a cold current—as at the left flank of the Gulf Stream in the neighbourhood of the Grand Banks off Newfoundland and at the subtropical and Antarctic convergences in the oceans of the Southern Hemisphere—the strong thermal gradients in the sea surface result in marked differences in the heating of the atmosphere on either side of the boundary. These temperature gradients tend to position and guide the strongest flow of the jet stream in the atmosphere above and thereby influence the development and steering of weather systems.

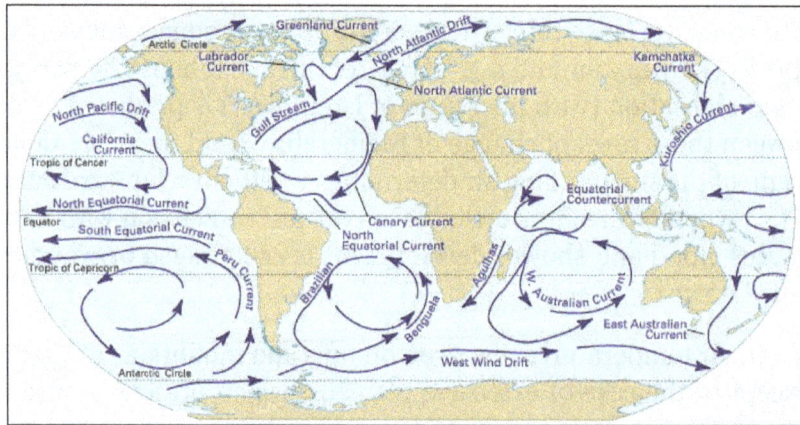

Major surface currents of the world's oceans. Subsurface currents also
move vast amounts of water.

Interactions between the ocean and the atmosphere proceed in both directions. They also operate at different rates. Some interesting lag effects, which are of value in long-range weather forecasting, arise through the considerably slower circulation of the ocean. Thus, enhanced strength of the easterly trade winds over low latitudes of the Atlantic north and south of the Equator impels more water toward the Caribbean and the Gulf of Mexico, producing a stronger flow and greater warmth in the Gulf Stream approximately six months later. Anomalies in the position of the Gulf Stream–Labrador Current boundary, which produce a greater or lesser extent of warm water near the Grand Banks, so affect the energy supply to the atmosphere and the development and steering of weather systems from that region that they are associated with rather persistent anomalies of weather pattern over the British Isles and northern Europe. Anomalies in the equatorial Pacific and in the northern limit of the Kuroshio Current (also called the Japan Current) seem to have effects on a similar scale. Indeed, through their influence on the latitude of the jet stream and the wavelength (that is, the spacing of cold trough and warm ridge regions) in the upper westerlies, these ocean anomalies exercise an influence over the atmospheric circulation that spreads to all parts of the hemisphere.

Sea-surface temperature anomalies that recur in the equatorial Pacific at variable intervals of two to seven years can sometimes produce major climatic perturbations. One such anomaly is known as El Niño (Spanish for "The Child"; it was so named by Peruvian fishermen who noticed its onset during the Christmas season).

During an El Niño event, warm surface water flows eastward from the equatorial Pacific, in at least partial response to weakening of the equatorial easterly winds, and replaces the normally cold up-welling surface water off the coast of Peru and Ecuador that is associated with the northward propagation of the cold Peru Current. The change in sea-surface temperature transforms the coastal climate from arid to wet. The event also affects atmospheric circulation in both hemispheres and is associated with changes in precipitation in regions of North America, Africa, and the western Pacific.

Short-term Temperature Changes

Many interesting short-term temperature fluctuations also occur, usually in connection with local weather disturbances. The rapid passage of a mid-latitude cold front, for example, can drop temperatures by 10 °C (18 °F) in a few minutes and, if followed by the sustained movement of a cold air mass, by as much as 50 °C in 24 hours, with life-threatening implications for the unwary. Temperature increases of up to 40 °C in a few hours also are possible downwind of major mountain ranges when air that has been warmed by the release of latent heat on the windward side of a range is forced to descend rapidly on the other side (such a wind is variously called chinook, foehn, or Santa Ana).

Geography

Climate is the prevailing patterns of temperature and precipitation across a region. A region's climate can be tropical or frigid, rainy or arid, temperate or monsoonal. Geography, or location, is one of the major determining factors in climate across the globe. Geography itself can be divided into components including distance from the equator, elevation above sea level, distance from water and topography, or the relief of the landscape.

Higher Latitudes Have Cooler Climates

Latitude is a measure of distance from the equator. Locations between the Tropic of Cancer and the Tropic of Capricorn, between 23 degrees north and 23 degrees south latitude, are considered tropical. As you move away from the equator, climates shift incrementally through subtropical, temperate, subarctic and, finally, arctic at the poles. The tilt of the Earth on its axis means that the further you get from the equator, the longer the area spends tilted away from the sun each year, and the cooler and more seasonal the climate.

Water Bodies Regulate Precipitation and Moderate Climate

Over 70 percent of the Earth's surface is covered in water, so it makes sense that water bodies influence climate. Oceans and lakes are very good at storing the heat that is created when the sun's energy is absorbed by the water. The water heats and adds moisture to the air above it, a process that drives the major air currents around the world. Water bodies also make the climate of adjacent land masses more moderate. They absorb extra heat during warm periods and release it during cooler periods. Warm, moist ocean air drives precipitation patterns around the world when it falls as precipitation as it is carried over cooler land masses.

Mountains Disrupt Air Flow

Mountain ranges are barriers to the smooth movement of air currents across continents. When an air mass encounters mountains, it is slowed and cooled because the air is forced up into cooler parts of the atmosphere in order to move over the obstruction. The cooled air can no longer hold as much moisture and releases it as precipitation on the mountain range. Once the air is over the mountain, it no longer has much moisture, and the leeward side of mountain ranges is drier than the windward side.

Higher Elevations Have Cooler Climates

Climates become cooler and the cold season lasts longer as elevation above sea level rises. This holds true for mountains and high-elevation plateaus, such as the steppes of Mongolia. Every 1.61 kilometers (1 mile) in elevation gain is roughly equivalent to moving 1,290 kilometers (800 miles) further from the equator. Mechanistically, higher elevations have lower air pressure, fewer atoms per unit of air to excite and thus, cooler temperatures. Mountains frequently receive more precipitation than the surrounding lowlands, but many high-altitude plains are deserts because of their location on the leeward side of a mountain range or continental mass.

Latitude

The tilt is the primary reason that different latitudes experience different weather patterns or climates. Outer planets, such as Saturn, have similar tilts, but they don't experience latitude-dependent climate variations in the same way because they aren't as close to the sun.

Any point on the surface of the Earth can be defined by a pair of angular coordinates known as longitude and latitude. Longitude is a line stretching from pole to pole with a given angular displacement from the Prime Meridian, which runs through Greenwich, England. Latitude is defined as the angular distance from the equator and is designated North or South depending on the hemisphere. The equator defines zero degrees latitude, which locates the North and South Poles at 90 degrees North and South respectively.

Temperatures Cool with Increasing Latitude

As latitude increases, the sun shines more obliquely and provides less warming energy. The equator always faces the sun directly, so the climate is warm year-round, with the average day and night temperature hovering between 12.5 and 14.3 degrees Celsius (54.5 and 57.7 degrees Fahrenheit). At the poles, however, winter and summer temperatures show a wider variation. The average temperature in the Arctic varies from zero °C (32 °F) in summer to -40 °C (-40 °F) in winter, while in the Antarctic, the temperature varies from -28.2 °C (-18 °F) in summer to -60 °C (-76 °F) in winter. The Antarctic is colder for two reasons: it's a landmass, and it's at a higher elevation than the Arctic.

The Earth's tilt affects the angle of incident sunlight on a particular location, but if that were its only effect, you would expect higher temperatures at each pole in summer. After all, that's when the pole is facing the sun and is actually slightly closer to it than the equator. This doesn't happen because at other times of the year the sun's rays have to pass through a thicker atmospheric filter than at the equator, producing cold enough temperatures to create permanent ice. In the summer, some of this ice melts, but the ice that doesn't melt reflects sunlight and prevents it from warming the atmosphere to the same extent it does at the equator.

Three Climatic Zones

Average temperatures cool with increasing latitude, producing well-defined climatic zones on the planet:

- The Tropic Zones extend from the equator north to the Tropic of Cancer at 23.5 degrees north to the Tropic of Capricorn at 23.5 degrees south. This is a region of generally warm temperatures and lush tropical vegetation.

- The Temperate Zones extend from the Tropics of Cancer and Capricorn to the Arctic and Antarctic Circles, which are located at 66.5 degrees north and south latitude respectively. These regions experience moderate temperatures and large temperature variations. The summers are hot and the winters cool.

- The Polar Zones extend from the Arctic and Antarctic Circles to the poles. In these regions, temperatures are cold and vegetation sparse.

Neutral Phase

During ENSO (El Niño Southern Oscillation) neutral conditions, surface trade winds blow westward across the equatorial Pacific Ocean. Blowing against the ocean's surface, these winds result in a westward current.

These persistent winds and currents are responsible for sea levels in the western Pacific, for example, in Indonesia, to be up to 50 cm higher than those in the eastern Pacific, for example, in Peru.

Neutral conditions also foster upwelling of cool, nutrient-rich sea water on the northern Pacific coast of South America. This process supports a healthy marine ecosystem and promotes robust aquaculture activities. Compared to other equatorial Pacific regions, the equatorial region from South America to the central Pacific will experience relatively cool sea temperatures contributing to relatively dry conditions in that area.

During ENSO Neutral, the following conditions are likely:

- Lower Atmospheric Circulation: Towards the West - Surface air pressure is higher over the central equatorial Pacific, for example Tahiti, than the western equatorial Pacific, for example Australia. This gradient in surface air pressure results in a flow of surface air moving from higher pressure to lower pressure, or west to east.

- Surface Water Circulation: Towards the West - The lower atmospheric circulation has a direct effect on the surface water beneath it. The east to west movement of air in the lower atmosphere along the equatorial Pacific Ocean works on the water below it, resulting in an east to west current near the surface.

- Upper Atmospheric Circulation: Towards the East - The low level winds converge over the warm waters of the western Pacific and rise. The forced ascent of the very moist air creates heavy rainfall in the region, wringing the air of moisture. The now dry air diverges out of the top of the convective region, moving out over the eastern Pacific to sink over the cooler waters.

- Thermocline: Slanted down from the eastern Pacific to the western Pacific - The thermocline marks the transition between the warm upper water and the cold deep water in the Pacific Ocean. Under neutral conditions, the thermocline is slanted down from east to west across the equatorial Pacific Ocean. This allows for upwelling of cooler, nutrient rich deep water towards the surface layer in the eastern equatorial Pacific.

References

- Atmospheric-humidity-and-precipitation, climate-meteorology: britannica.com, Retrieved 15 January, 2019

- AtmosCirculation: climate.ncsu.edu, Retrieved 16 April, 2019

- Do-currents-affect-weather-climate-7735765: sciencing.com, Retrieved 18 March, 2019

- Surface-currents: courses.lumenlearning.com, Retrieved 16 June, 2019

- Large-bodies-water-affect-climate-coastal-areas-22337: sciencing.com, Retrieved 27 April, 2019

- Latitude-affect-climate-4586935: sciencing.com, Retrieved 16 June, 2019

- Phase-neutral: iridl.ldeo.columbia.edu, Retrieved 26 March, 2019

Chapter 4

Climate Change and Global Warming

Long-term rise in the average temperature of the Earth's climate system is referred to as global warming. It consists of various studies such as impacts of climate change on ecosystems and human health, air pollution and climate change, etc. Climate change and global warming is an interdisciplinary subject which makes it essential to understand its related studies.

Climate Change

Climate change is the periodic modification of Earth's climate brought about as a result of changes in the atmosphere as well as interactions between the atmosphere and various other geologic, chemical, biological, and geographic factors within the Earth system.

A series of photographs of the Grinnell Glacier taken from the
summit of Mount Gould in Glacier National Park, Montana, (from left to right).

The atmosphere is a dynamic fluid that is continually in motion. Both its physical properties and its rate and direction of motion are influenced by a variety of factors, including solar radiation, the geographic position of continents, ocean currents, the location and orientation of mountain ranges, atmospheric chemistry, and vegetation growing on the land surface. All these factors change through time. Some factors, such as the distribution of heat within the oceans, atmospheric chemistry, and surface vegetation, change at very short timescales. Others, such as the position of continents and the location and height of mountain ranges, change over very long timescales. Therefore, climate, which results from the physical properties and motion of the atmosphere, varies at every conceivable timescale.

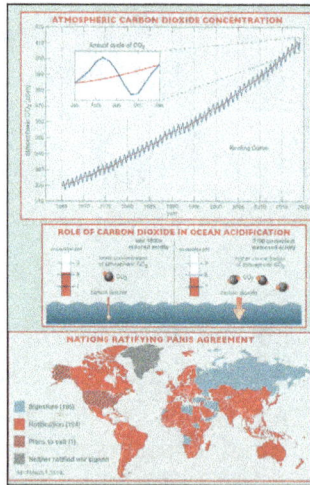

Climate change: timeline A timeline of important developments in climate change.

Climate is often defined loosely as the average weather at a particular place, incorporating such features as temperature, precipitation, humidity, and windiness. A more specific definition would state that climate is the mean state and variability of these features over some extended time period. Both definitions acknowledge that the weather is always changing, owing to instabilities in the atmosphere. And as weather varies from day to day, so too does climate vary, from daily day-and-night cycles up to periods of geologic time hundreds of millions of years long. In a very real sense, climate variation is a redundant expression—climate is always varying. No two years are exactly alike, nor are any two decades, any two centuries, or any two millennia.

Earth System

The atmosphere is influenced by and linked to other features of Earth, including oceans, ice masses (glaciers and sea ice), land surfaces, and vegetation. Together, they make up an integrated Earth system, in which all components interact with and influence one another in often complex ways. For instance, climate influences the distribution of vegetation on Earth's surface (e.g., deserts exist in arid regions, forests in humid regions), but vegetation in turn influences climate by reflecting radiant energy back into the atmosphere, transferring water (and latent heat) from soil to the atmosphere, and influencing the horizontal movement of air across the land surface.

Iceberg Tourist boat in front of a massive iceberg near the coast of Greenland.

Earth scientists and atmospheric scientists are still seeking a full understanding of the complex feedbacks and interactions among the various components of the Earth system. This effort is being facilitated by the development of an interdisciplinary science called Earth system science. Earth system science is composed of a wide range of disciplines, including climatology (the study of the atmosphere), geology (the study of Earth's surface and underground processes), ecology (the study of how Earth's organisms relate to one another and their environment), oceanography (the study of Earth's oceans), glaciology (the study of Earth's ice masses), and even the social sciences (the study of human behaviour in its social and cultural aspects).

A full understanding of the Earth system requires knowledge of how the system and its components have changed through time. The pursuit of this understanding has led to development of Earth system history, an interdisciplinary science that includes not only the contributions of Earth system scientists but also paleontologists (who study the life of past geologic periods), paleoclimatologists (who study past climates), paleoecologists (who study past environments and ecosystems), paleoceanographers (who study the history of the oceans), and other scientists concerned with Earth history. Because different components of the Earth system change at different rates and are relevant at different timescales, Earth system history is a diverse and complex science. Students of Earth system history are not just concerned with documenting what has happened; they also view the past as a series of experiments in which solar radiation, ocean currents, continental configurations, atmospheric chemistry, and other important features have varied. These experiments provide opportunities to learn the relative influences of and interactions between various components of the Earth system. Studies of Earth system history also specify the full array of states the system has experienced in the past and those the system is capable of experiencing in the future.

Undoubtedly, people have always been aware of climatic variation at the relatively short timescales of seasons, years, and decades. Biblical scripture and other early documents refer to droughts, floods, periods of severe cold, and other climatic events. Nevertheless, a full appreciation of the nature and magnitude of climatic change did not come about until the late 18th and early 19th centuries, a time when the widespread recognition of the deep antiquity of Earth occurred. Naturalists of this time, including Scottish geologist Charles Lyell, Swiss-born naturalist and geologist Louis Agassiz, English naturalist Charles Darwin, American botanist Asa Gray, and Welsh naturalist Alfred Russel Wallace, came to recognize geologic and biogeographic evidence that made sense only in the light of past climates radically different from those prevailing today.

The occurrence of multiple epochs in recent Earth history during which continental glaciers, developed at high latitudes, penetrated into northern Europe and eastern North America was recognized by scientists by the late 19th century. Scottish geologist James Croll proposed that recurring variations in orbital eccentricity (the deviation of Earth's orbit from a perfectly circular path) were responsible for alternating glacial and interglacial periods. Croll's controversial idea was taken up by Serbian mathematician and astronomer Milutin Milankovitch in the early 20th century. Milankovitch proposed that the mechanism that brought about periods of glaciation was driven by cyclic changes in eccentricity as well as two other orbital parameters: precession (a change in the directional focus of Earth's axis of rotation) and axial tilt (a change in the inclination of Earth's axis with respect to the plane of its orbit around the Sun). Orbital variation is now recognized as an important driver of climatic variation throughout Earth's history.

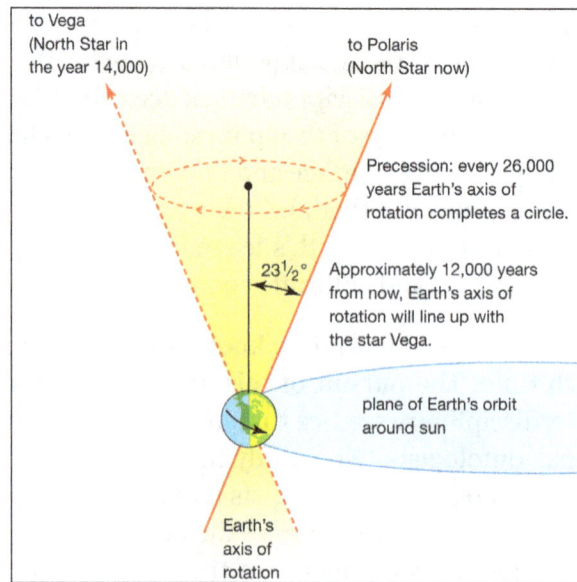

The precession of Earth's axis.

Evidence for Climate Change

All historical sciences share a problem: As they probe farther back in time, they become more reliant on fragmentary and indirect evidence. Earth system history is no exception. High-quality instrumental records spanning the past century exist for most parts of the world, but the records become sparse in the 19th century, and few records predate the late 18th century. Other historical documents, including ship's logs, diaries, court and church records, and tax rolls, can sometimes be used. Within strict geographic contexts, these sources can provide information on frosts, droughts, floods, sea ice, the dates of monsoons, and other climatic features—in some cases up to several hundred years ago.

Fortunately, climatic change also leaves a variety of signatures in the natural world. Climate influences the growth of trees and corals, the abundance and geographic distribution of plant and animal species, the chemistry of oceans and lakes, the accumulation of ice in cold regions, and the erosion and deposition of materials on Earth's surface. Paleoclimatologists study the traces of these effects, devising clever and subtle ways to obtain information about past climates. Most of the evidence of past climatic change is circumstantial, so paleoclimatology involves a great deal of investigative work. Wherever possible, paleoclimatologists try to use multiple lines of evidence to cross-check their conclusions. They are frequently confronted with conflicting evidence, but this, as in other sciences, usually leads to an enhanced understanding of the Earth system and its complex history. New sources of data, analytical tools, and instruments are becoming available, and the field is moving quickly. Revolutionary changes in the understanding of Earth's climate history have occurred since the 1990s, and coming decades will bring many new insights and interpretations.

Ongoing climatic changes are being monitored by networks of sensors in space, on the land surface, and both on and below the surface of the world's oceans. Climatic changes of the past 200–300 years, especially since the early 1900s, are documented by instrumental records and other archives. These written documents and records provide information about climate change in some locations for the past few hundred years. Some very rare records date back over 1,000 years. Researchers studying climatic changes predating the instrumental record rely increasingly on natural archives, which are biological or geologic processes that record some aspect of past climate. These natural archives, often referred to as proxy evidence, are extraordinarily diverse; they include, but are not limited to, fossil records of past plant and animal distributions, sedimentary and geochemical indicators of former conditions of oceans and continents, and land surface features characteristic of past climates. Paleoclimatologists study these natural archives by collecting cores, or cylindrical samples, of sediments from lakes, bogs, and oceans; by studying surface features and geological strata; by examining tree ring patterns from cores or sections of living and dead trees; by drilling into marine corals and cave stalagmites; by drilling into the ice sheets of Antarctica and Greenland and the high-elevation glaciers of the Plateau of Tibet, the Andes, and other montane regions; and by a wide variety of other means. Techniques for extracting paleoclimatic information are continually being developed and refined, and new kinds of natural archives are being recognized and exploited.

Causes of Climate Change

It is much easier to document the evidence of climate variability and past climate change than it is to determine their underlying mechanisms. Climate is influenced by a multitude of factors that operate at timescales ranging from hours to hundreds of millions of years. Many of the causes of climate change are external to the Earth system. Others are part of the Earth system but external to the atmosphere. Still others involve interactions between the atmosphere and other components of the Earth system and are collectively described as feedbacks within the Earth system. Feedbacks are among the most recently discovered and challenging causal factors to study. Nevertheless, these factors are increasingly recognized as playing fundamental roles in climate variation.

Solar Variability

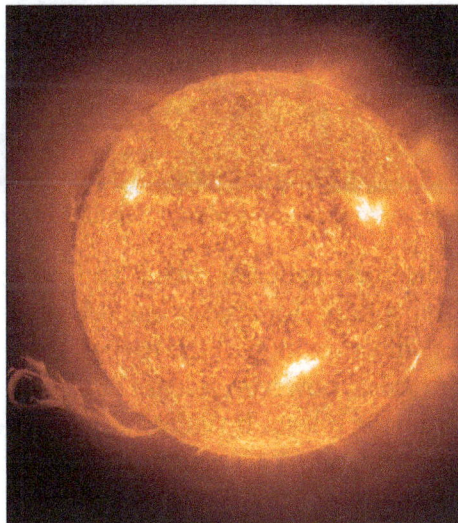

The Sun as imaged in extreme ultraviolet light by the Earth-orbiting Solar and Heliospheric Observatory (SOHO) satellite. A massive loop-shaped eruptive prominence is visible at the lower left. Nearly white areas are the hottest; deeper reds indicate cooler temperatures.

The luminosity, or brightness, of the Sun has been increasing steadily since its formation. This phenomenon is important to Earth's climate, because the Sun provides the energy to drive atmospheric circulation and constitutes the input for Earth's heat budget. Low solar luminosity during Precambrian time underlies the faint young Sun paradox,.

Radiative energy from the Sun is variable at very small timescales, owing to solar storms and other disturbances, but variations in solar activity, particularly the frequency of sunspots, are also documented at decadal to millennial timescales and probably occur at longer timescales as well. The "Maunder minimum," a period of drastically reduced sunspot activity between AD 1645 and 1715, has been suggested as a contributing factor to the Little Ice Age.

Volcanic Activity

Volcanic activity can influence climate in a number of ways at different timescales. Individual volcanic eruptions can release large quantities of sulfur dioxide and other aerosols into the stratosphere, reducing atmospheric transparency and thus the amount of solar radiation reaching Earth's surface and troposphere. A recent example is the 1991 eruption in the Philippines of Mount Pinatubo, which had measurable influences on atmospheric circulation and heat budgets. The 1815 eruption of Mount Tambora on the island of Sumbawa had more dramatic consequences, as the spring and summer of the following year (1816, known as "the year without a summer") were unusually cold over much of the world. New England and Europe experienced snowfalls and frosts throughout the summer of 1816.

A column of gas and ash rising from Mount Pinatubo in the Philippines

Volcanoes and related phenomena, such as ocean rifting and subduction, release carbon dioxide into both the oceans and the atmosphere. Emissions are low; even a massive volcanic eruption such as Mount Pinatubo releases only a fraction of the carbon dioxide emitted by fossil-fuel combustion in a year. At geologic timescales, however, release of this greenhouse gas can have important effects. Variations in carbon dioxide release by volcanoes and ocean rifts over millions of years can alter the chemistry of the atmosphere. Such changeability in carbon dioxide concentrations probably accounts for much of the climatic variation that has taken place during the Phanerozoic Eon.

Tectonic Activity

Tectonic movements of Earth's crust have had profound effects on climate at timescales of millions to tens of millions of years. These movements have changed the shape, size, position, and elevation of the continental masses as well as the bathymetry of the oceans. Topographic and bathymetric changes in turn have had strong effects on the circulation of both the atmosphere and the oceans. For example, the uplift of the Tibetan Plateau during the Cenozoic Era affected atmospheric circulation patterns, creating the South Asian monsoon and influencing climate over much of the rest of Asia and neighbouring regions.

Tectonic activity also influences atmospheric chemistry, particularly carbon dioxide concentrations. Carbon dioxide is emitted from volcanoes and vents in rift zones and subduction zones. Variations in the rate of spreading in rift zones and the degree of volcanic activity near plate margins have influenced atmospheric carbon dioxide concentrations throughout Earth's history. Even the chemical weathering of rock constitutes an important sink for carbon dioxide. (A carbon sink is any process that removes carbon dioxide from the atmosphere by the chemical conversion of CO_2 to organic or inorganic carbon compounds). Carbonic acid, formed from carbon dioxide and water, is a reactant in dissolution of silicates and other minerals. Weathering rates are related to the mass, elevation, and exposure of bedrock. Tectonic uplift can increase all these factors and thus lead to increased weathering and carbon dioxide absorption. For example, the chemical weathering of the rising Tibetan Plateau may have played an important role in depleting the atmosphere of carbon dioxide during a global cooling period in the late Cenozoic Era.

Orbital (Milankovich) Variations

The orbital geometry of Earth is affected in predictable ways by the gravitational influences of other planets in the solar system. Three primary features of Earth's orbit are affected, each in a cyclic, or regularly recurring, manner. First, the shape of Earth's orbit around the Sun, varies from nearly circular to elliptical (eccentric), with periodicities of 100,000 and 413,000 years. Second, the tilt of Earth's axis with respect to the Sun, which is primarily responsible for Earth's seasonal climates, varies between 22.1° and 24.5° from the plane of Earth's rotation around the Sun. This variation occurs on a cycle of 41,000 years. In general, the greater the tilt, the greater the solar radiation received by hemispheres in summer and the less received in winter. The third cyclic change to Earth's orbital geometry results from two combined phenomena: (1) Earth's axis of rotation wobbles, changing the direction of the axis with respect to the Sun, and (2) the orientation of Earth's orbital ellipse rotates slowly. These two processes create a 26,000-year cycle, called precession of the equinoxes, in which the position of Earth at the equinoxes and solstices changes. Today Earth is closest to the Sun (perihelion) near the December solstice, whereas 9,000 years ago perihelion occurred near the June solstice.

These orbital variations cause changes in the latitudinal and seasonal distribution of solar radiation, which in turn drive a number of climate variations. Orbital variations play major roles in pacing glacial-interglacial and monsoonal patterns. Their influences have been identified in climatic changes over much of the Phanerozoic. For example, cyclothems—which are interbedded marine, fluvial, and coal beds characteristic of the Pennsylvanian Subperiod (318.1 million to 299 million years ago)—appear to represent Milankovitch-driven changes in mean sea level.

Greenhouse Gases

Greenhouse gases are gas molecules that have the property of absorbing infrared radiation (net heat energy) emitted from Earth's surface and reradiating it back to Earth's surface, thus contributing to the phenomenon known as the greenhouse effect. Carbon dioxide, methane, and water vapour are the most important greenhouse gases, and they have a profound effect on the energy budget of the Earth system despite making up only a fraction of all atmospheric gases. Concentrations of greenhouse gases have varied substantially during Earth's history, and these variations have driven substantial climate changes at a wide range of timescales. In general, greenhouse gas concentrations have been particularly high during warm periods and low during cold phases. A number of processes influence greenhouse gas concentrations. Some, such as tectonic activities, operate at timescales of millions of years, whereas others, such as vegetation, soil, wetland, and ocean sources and sinks, operate at timescales of hundreds to thousands of years. Human activities—especially fossil-fuel combustion since the Industrial Revolution—are responsible for steady increases in atmospheric concentrations of various greenhouse gases, especially carbon dioxide, methane, ozone, and chlorofluorocarbons (CFCs).

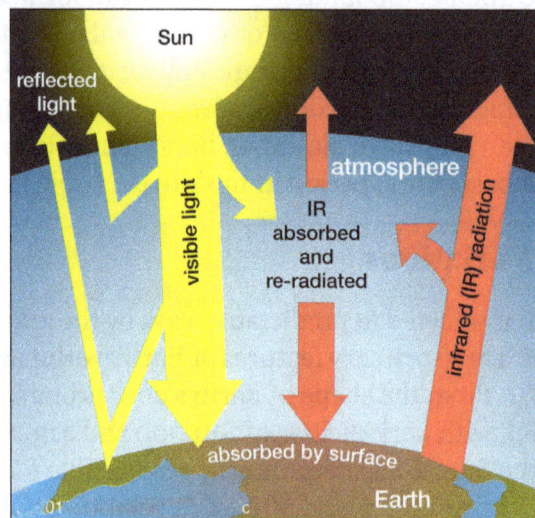

Greenhouse effect on earth.

The greenhouse effect on Earth. Some incoming sunlight is reflected by Earth's atmosphere and surface, but most is absorbed by the surface, which is warmed. Infrared (IR) radiation is then emitted from the surface. Some IR radiation escapes to space, but some is absorbed by the atmosphere's greenhouse gases (especially water vapour, carbon dioxide, and methane) and reradiated in all directions, some to space and some back toward the surface, where it further warms the surface and the lower atmosphere.

Feedback

The feedbacks involve different components that operate at different rates and timescales. Ice sheets, sea ice, terrestrial vegetation, ocean temperatures, weathering rates, ocean circulation, and greenhouse gas concentrations are all influenced either directly or indirectly by the atmosphere; however, they also all feed back into the atmosphere, thereby influencing it in important ways. For example, different forms and densities of vegetation on the land surface influence the albedo, or reflectivity, of Earth's surface, thus affecting the overall radiation budget at local to regional scales. At the same time, the transfer of water molecules from soil to the atmosphere is mediated by vegetation, both directly (from transpiration through plant stomata) and indirectly (from shading and temperature influences on direct evaporation from soil). This regulation of latent heat flux by vegetation can influence climate at local to global scales. As a result, changes in vegetation, which are partially controlled by climate, can in turn influence the climate system. Vegetation also influences greenhouse gas concentrations; living plants constitute an important sink for atmospheric carbon dioxide, whereas they act as sources of carbon dioxide when they are burned by wildfires or undergo decomposition. These and other feedbacks among the various components of the Earth system are critical for both understanding past climate changes and predicting future ones.

Mixed evergreen and hardwood forest on the slopes of the Adirondack Mountains near Keene Valley, New York.

Human Activities

Recognition of global climate change as an environmental issue has drawn attention to the climatic impact of human activities. Most of this attention has focused on carbon dioxide emission via fossil-fuel combustion and deforestation. Human activities also yield releases of other greenhouse gases, such as methane (from rice cultivation, livestock, landfills, and other sources) and chlorofluorocarbons (from industrial sources). There is little doubt among climatologists that these greenhouse gases affect the radiation budget of Earth; the nature and magnitude of the climatic response are a subject of intense research activity. Paleoclimate records from tree rings, coral, and ice cores indicate a clear warming trend spanning the entire 20th century and the first decade of the 21st century. In fact, the 20th century was the warmest of the past 10 centuries, and the decade 2001–10 was the warmest decade since the beginning of modern instrumental record keeping. Many climatologists have pointed to this warming pattern as clear evidence of human-induced climate change resulting from the production of greenhouse gases.

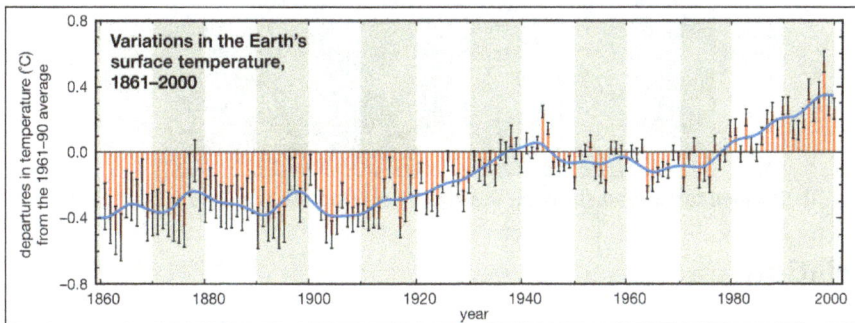

The global average surface temperature range for each year from 1861 to 2000 is shown by solid red bars, with the confidence range in the data for each year shown by thin whisker bars. The average change over time is shown by the solid curve.

A second type of human impact, the conversion of vegetation by deforestation, afforestation, and agriculture, is receiving mounting attention as a further source of climate change. It is becoming increasingly clear that human impacts on vegetation cover can have local, regional, and even global effects on climate, due to changes in the sensible and latent heat flux to the atmosphere and the distribution of energy within the climate system. The extent to which these factors contribute to recent and ongoing climate change is an important, emerging area of study.

Climate Change within a Human Life Span

Regardless of their locations on the planet, all humans experience climate variability and change within their lifetimes. The most familiar and predictable phenomena are the seasonal cycles, to which people adjust their clothing, outdoor activities, thermostats, and agricultural practices. However, no two summers or winters are exactly alike in the same place; some are warmer, wetter, or stormier than others. This interannual variation in climate is partly responsible for year-to-year variations in fuel prices, crop yields, road maintenance budgets, and wildfire hazards. Single-year, precipitation-driven floods can cause severe economic damage, such as those of the upper Mississippi River drainage basin during the summer of 1993, and loss of life, such as those that devastated much of Bangladesh in the summer of 1998. Similar damage and loss of life can also occur as the result of wildfires, severe storms, hurricanes, heat waves, and other climate-related events.

Climate variation and change may also occur over longer periods, such as decades. Some locations experience multiple years of drought, floods, or other harsh conditions. Such decadal variation of climate poses challenges to human activities and planning. For example, multiyear droughts can disrupt water supplies, induce crop failures, and cause economic and social dislocation, as in the case of the Dust Bowl droughts in the midcontinent of North America during the 1930s. Multiyear droughts may even cause widespread starvation, as in the Sahel drought that occurred in northern Africa during the 1970s and '80s.

Abandoned farmstead showing the effects of wind erosion in the Dust Bowl.

Seasonal Variation

Every place on Earth experiences seasonal variation in climate (though the shift can be slight in some tropical regions). This cyclic variation is driven by seasonal changes in the supply of solar radiation to Earth's atmosphere and surface. Earth's orbit around the Sun is elliptical; it is closer to the Sun (147 million km [about 91 million miles]) near the winter solstice and farther from the Sun (152 million km [about 94 million miles]) near the summer solstice in the Northern Hemisphere. Furthermore, Earth's axis of rotation occurs at an oblique angle (23.5°) with respect to its orbit. Thus, each hemisphere is tilted away from the Sun during its winter period and toward the Sun in its summer period. When a hemisphere is tilted away from the Sun, it receives less solar radiation than the opposite hemisphere, which at that time is pointed toward the Sun. Thus, despite the closer proximity of the Sun at the winter solstice, the Northern Hemisphere receives less solar radiation during the winter than it does during the summer. Also as a consequence of the tilt, when the Northern Hemisphere experiences winter, the Southern Hemisphere experiences summer.

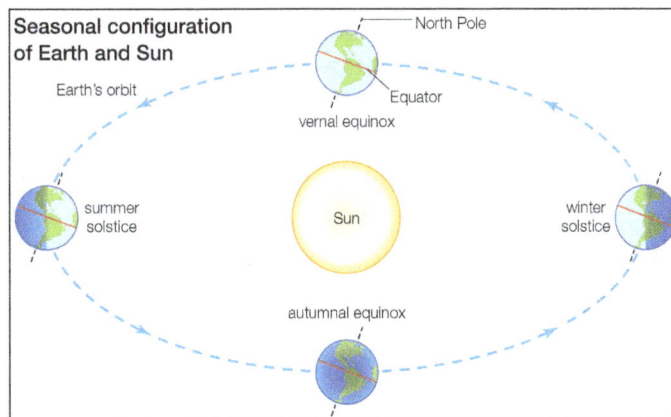

A diagram shows the position of Earth at the beginning of each season in the Northern Hemisphere.

Earth's climate system is driven by solar radiation; seasonal differences in climate ultimately result from the seasonal changes in Earth's orbit. The circulation of air in the atmosphere and water in the oceans responds to seasonal variations of available energy from the Sun. Specific seasonal changes in climate occurring at any given location on Earth's surface largely result from the transfer of energy from atmospheric and oceanic circulation. Differences in surface heating taking place between summer and winter cause storm tracks and pressure centres to shift position and strength. These heating differences also drive seasonal changes in cloudiness, precipitation, and wind.

Seasonal responses of the biosphere (especially vegetation) and cryosphere (glaciers, sea ice, snowfields) also feed into atmospheric circulation and climate. Leaf fall by deciduous trees as they go into winter dormancy increases the albedo (reflectivity) of Earth's surface and may lead to greater local and regional cooling. Similarly, snow accumulation also increases the albedo of land surfaces and often amplifies winter's effects.

Interannual Variation

Interannual climate variations, including droughts, floods, and other events, are caused by a complex array of factors and Earth system interactions. One important feature that plays a role in these variations is the periodic change of atmospheric and oceanic circulation patterns in the tropical Pacific region, collectively known as El Niño–Southern Oscillation (ENSO) variation. Although its primary climatic effects are concentrated in the tropical Pacific, ENSO has cascading effects that often extend to the Atlantic Ocean region, the interior of Europe and Asia, and the polar regions. These effects, called teleconnections, occur because alterations in low-latitude atmospheric circulation patterns in the Pacific region influence atmospheric circulation in adjacent and downstream systems. As a result, storm tracks are diverted and atmospheric pressure ridges (areas of high pressure) and troughs (areas of low pressure) are displaced from their usual patterns.

As an example, El Niño events occur when the easterly trade winds in the tropical Pacific weaken or reverse direction. This shuts down the upwelling of deep, cold waters off the west coast of South America, warms the eastern Pacific, and reverses the atmospheric pressure gradient in the western Pacific. As a result, air at the surface moves eastward from Australia and Indonesia toward the central Pacific and the Americas. These changes produce high rainfall and flash floods along the normally arid coast of Peru and severe drought in the normally wet regions of northern Australia and Indonesia. Particularly severe El Niño events lead to monsoon failure in the Indian Ocean region, resulting in intense drought in India and East Africa. At the same time, the westerlies and storm tracks are displaced toward the Equator, providing California and the desert Southwest of the United States with wet, stormy winter weather and causing winter conditions in the Pacific Northwest, which are typically wet, to become warmer and drier. Displacement of the westerlies also results in drought in northern China and from northeastern Brazil through sections of Venezuela. Long-term records of ENSO variation from historical documents, tree rings, and reef corals indicate that El Niño events occur, on average, every two to seven years. However, the frequency and intensity of these events vary through time.

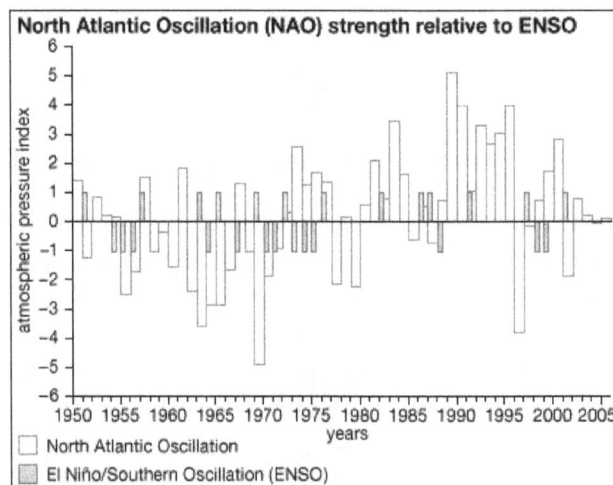

The North Atlantic Oscillation (NAO) is another example of an interannual oscillation that produces important climatic effects within the Earth system and can influence climate throughout the Northern Hemisphere. This phenomenon results from variation in the pressure gradient, or the difference in atmospheric pressure between the subtropical high, usually situated between the Azores and Gibraltar, and the Icelandic low, centred between Iceland and Greenland. When the pressure gradient is steep due to a strong subtropical high and a deep Icelandic low (positive phase), northern Europe and northern Asia experience warm, wet winters with frequent strong winter storms. At the same time, southern Europe is dry. The eastern United States also experiences warmer, less snowy winters during positive NAO phases, although the effect is not as great as in Europe. The pressure gradient is dampened when NAO is in a negative mode—that is, when a weaker pressure gradient exists from the presence of a weak subtropical high and Icelandic low. When this happens, the Mediterranean region receives abundant winter rainfall, while northern Europe is cold and dry. The eastern United States is typically colder and snowier during a negative NAO phase.

During years when the North Atlantic Oscillation (NAO) is in its positive phase, the eastern United

States, southeastern Canada, and northwestern Europe experience warmer winter temperatures, whereas colder temperatures are found in these locations during its negative phase. When the El Niño/Southern Oscillation (ENSO) and NAO are both in their positive phase, European winters tend to be wetter and less severe; however, beyond this general tendency, the influence of the ENSO upon the NAO is not well understood.

The ENSO and NAO cycles are driven by feedbacks and interactions between the oceans and atmosphere. Interannual climate variation is driven by these and other cycles, interactions among cycles, and perturbations in the Earth system, such as those resulting from large injections of aerosols from volcanic eruptions. One example of a perturbation due to volcanism is the 1991 eruption of Mount Pinatubo in the Philippines, which led to a decrease in the average global temperature of approximately 0.5 °C (0.9 °F) the following summer.

Decadal Variation

Climate varies on decadal timescales, with multiyear clusters of wet, dry, cool, or warm conditions. These multiyear clusters can have dramatic effects on human activities and welfare. For instance, a severe three-year drought in the late 16th century probably contributed to the destruction of Sir Walter Raleigh's "Lost Colony" at Roanoke Island in what is now North Carolina, and a subsequent seven-year drought led to high mortality at the Jamestown Colony in Virginia. Also, some scholars have implicated persistent and severe droughts as the main reason for the collapse of the Maya civilization in Mesoamerica between AD 750 and 950; however, discoveries in the early 21st century suggest that war-related trade disruptions played a role, possibly interacting with famines and other drought-related stresses.

Although decadal-scale climate variation is well documented, the causes are not entirely clear. Much decadal variation in climate is related to interannual variations. For example, the frequency and magnitude of ENSO change through time. The early 1990s were characterized by repeated El Niño events, and several such clusters have been identified as having taken place during the 20th century. The steepness of the NAO gradient also changes at decadal timescales; it has been particularly steep since the 1970s.

Recent research has revealed that decadal-scale variations in climate result from interactions between the ocean and the atmosphere. One such variation is the Pacific Decadal Oscillation (PDO), also referred to as the Pacific Decadal Variability (PDV), which involves changing sea surface temperatures (SSTs) in the North Pacific Ocean. The SSTs influence the strength and position of the Aleutian Low, which in turn strongly affects precipitation patterns along the Pacific Coast of North America. PDO variation consists of an alternation between "cool-phase" periods, when coastal Alaska is relatively dry and the Pacific Northwest relatively wet, and "warm-phase" periods, characterized by relatively high precipitation in coastal Alaska and low precipitation in the Pacific Northwest. Tree ring and coral records, which span at least the last four centuries, document PDO variation.

A similar oscillation, the Atlantic Multidecadal Oscillation (AMO), occurs in the North Atlantic and strongly influences precipitation patterns in eastern and central North America. A warm-phase AMO (relatively warm North Atlantic SSTs) is associated with relatively high rainfall in Florida and low rainfall in much of the Ohio Valley. However, the AMO interacts with the PDO, and both interact with interannual variations, such as ENSO and NAO, in complex ways. Such interactions may lead to the amplification of droughts, floods, or other climatic anomalies. For example, severe droughts over much of the conterminous United States in the first few years of the 21st century were associated with warm-phase AMO combined with cool-phase PDO. The mechanisms underlying decadal variations, such as PDO and AMO, are poorly understood, but they are probably related to ocean-atmosphere interactions with larger time constants than interannual variations. Decadal climatic variations are the subject of intense study by climatologists and paleoclimatologists.

Impacts of Climate Change on Ecosystems

Climate is an important environmental influence on ecosystems. Changing climate affects ecosystems in a variety of ways. For instance, warming may force species to migrate to higher latitudes or higher elevations where temperatures are more conducive to their survival. Similarly, as sea level rises, saltwater intrusion into a freshwater system may force some key species to relocate or die, thus removing predators or prey that are critical in the existing food chain.

Climate change not only affects ecosystems and species directly, it also interacts with other human stressors such as development. Although some stressors cause only minor impacts when acting alone, their cumulative impact may lead to dramatic ecological changes. For instance, climate change may exacerbate the stress that land development places on fragile coastal areas. Additionally, recently logged forested areas may become vulnerable to erosion if climate change leads to increases in heavy rain storms.

Changes in the Timing of Seasonal Life Cycle Events

For many species, the climate where they live or spend part of the year influences key stages of their annual life cycle, such as migration, blooming, and reproduction. As winters have become shorter and milder, the timing of these events has changed in some parts of the country:

- Earlier springs have led to earlier nesting for 28 migratory bird species on the East Coast of the United States.

- Northeastern birds that winter in the southern United States are returning north in the spring 13 days earlier than they did in a century ago.

- In a California study, 16 out of 23 butterfly species shifted their migration timing and arrived earlier.

Because species differ in their ability to adjust, asynchronies can develop, increasing species and ecosystem vulnerability. These asynchronies can include mismatches in the timing of migration, breeding, pest avoidance, and food availability. Growth and survival are reduced when migrants arrive at a location before or after food sources are present.

Range Shifts

As temperatures increase, the habitat ranges of many North American species are moving north and to higher elevations. In recent decades, in both land and aquatic environments, plants and animals have moved to higher elevations at a median rate of 36 feet (0.011 kilometers) per decade, and to higher latitudes at a median rate of 10.5 miles (16.9 kilometers) per decade. While this means a range expansion for some species, for others it means movement into less hospitable habitat, increased competition, or range reduction, with some species having nowhere to go because they are already at the top of a mountain or at the northern limit of land suitable for their habitat. These factors lead to local extinctions of both plants and animals in some areas. As a result, the ranges of vegetative biomes are projected to change across 5-20% of the land in the United States by 2100.

For example, boreal forests are invading tundra, reducing habitat for the many unique species that depend on the tundra ecosystem, such as caribou, arctic foxes, and snowy owls. Other observed changes in the United States include a shift in the temperate broadleaf/conifer forest boundary in the Green Mountains of Vermont; a shift in the shrubland/conifer forest boundary in New Mexico; and an upward elevation shift of the temperate mixed/conifer forest boundary in Southern California.

As rivers and streams warm, warmwater fish are expanding into areas previously inhabited by coldwater species. As waters warm, coldwater fish, including many highly-valued trout and salmon species, are losing their habitat, with projections of 47% habitat loss by 2080. In certain regions in the western United States, losses of western trout populations may exceed 60 percent, while in other regions, losses of bull trout may reach about 90 percent. Range shifts disturb the current state of the ecosystem and can limit opportunities for fishing and hunting.

Food Web Disruptions

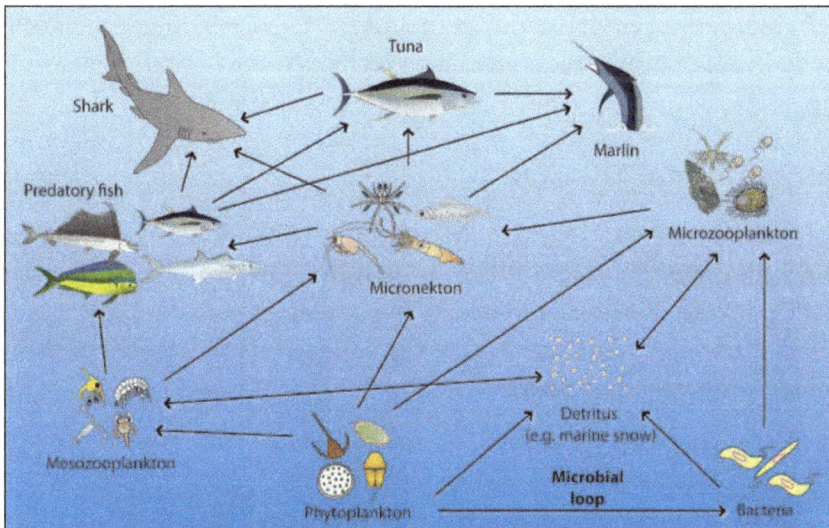

The Arctic food web is complex. The loss of sea ice can ultimately affect
the entire food web, from algae and plankton to fish to mammals.

The impact of climate change on a particular species can ripple through a food web and affect a wide range of other organisms. For example, the figure below shows the complex nature of the food web for polar bears. Not only is the decline of sea ice impairing polar bear populations by reducing the extent of their primary habitat, it is also negatively impacting them via food web effects.

Declines in the duration and extent of sea ice in the Arctic leads to declines in the abundance of ice algae, which thrive in nutrient-rich pockets in the ice. These algae are eaten by zooplankton, which are in turn eaten by Arctic cod, an important food source for many marine mammals, including seals. Seals are eaten by polar bears. Hence, declines in ice algae can contribute to declines in polar bear populations.

Buffer and Threshold Effects

Ecosystems can serve as natural buffers from extreme events such as wildfires, flooding, and drought. Climate change and human modification may restrict ecosystems' ability to temper the impacts of extreme conditions, and thus may increase vulnerability to damage. Examples include reefs and barrier islands that protect coastal ecosystems from storm surges, wetland ecosystems that absorb floodwaters, and cyclical wildfires that clear excess forest debris and reduce the risk of dangerously large fires.

In some cases, ecosystem change occurs rapidly and irreversibly because a threshold, or "tipping point," is passed. One area of concern for thresholds is the Prairie Pothole Region in the north-central part of the United States. This ecosystem is a vast area of small, shallow lakes, known as "prairie potholes" or "playa lakes." These wetlands provide essential breeding habitat for most North American waterfowl species. The pothole region has experienced temporary droughts in the past. However, a permanently warmer, drier future may lead to a threshold change—a dramatic drop in the prairie potholes that host waterfowl populations, which subsequently provide highly valued hunting and wildlife viewing opportunities.

Similarly, when coral reefs become stressed from increased ocean temperatures, they expel microorganisms that live within their tissues and are essential to their health. This is known as coral bleaching. As ocean temperatures warm and the acidity of the ocean increases, bleaching and coral die-offs are likely to become more frequent. Chronically stressed coral reefs are less likely to recover.

Pathogens, Parasites and Disease

Climate change and shifts in ecological conditions could support the spread of pathogens, parasites, and diseases, with potentially serious effects on human health, agriculture, and fisheries. For example, the oyster parasite, Perkinsus marinus, is capable of causing large oyster die-offs. This parasite has extended its range northward from Chesapeake Bay to Maine, a 310-mile expansion tied to above-average winter temperatures.

Extinction Risks

Climate change, along with habitat destruction and pollution, is one of the important stressors that can contribute to species extinction. The IPCC estimates that 20-30% of the plant and animal species evaluated so far in climate change studies are at risk of extinction if temperatures reach the levels projected to occur by the end of this century. Global rates of species extinctions are likely to approach or exceed the upper limit of observed natural rates of extinction in the fossil record. Examples of species that are particularly climate sensitive and could be at risk of significant losses include animals that are adapted to mountain environments, such as the pika; animals that are dependent on sea ice habitats, such as ringed seals and polar bears; and coldwater fish, such as salmon in the Pacific Northwest.

Impacts of Climate Change on Societal System

Climate change could affect our society through impacts on a number of different social, cultural, and natural resources. For example, climate change could affect human health, infrastructure, and transportation systems, as well as energy, food, and water supplies.

Some groups of people will likely face greater challenges than others. Climate change may especially impact people who live in areas that are vulnerable to coastal storms, drought, and sea level rise or people who live in poverty, older adults, and immigrant communities. Similarly, some types of professions and industries may face considerable challenges from climate change. Professions that are closely linked to weather and climate, such as outdoor tourism, commerce, and agriculture, will likely be especially affected.

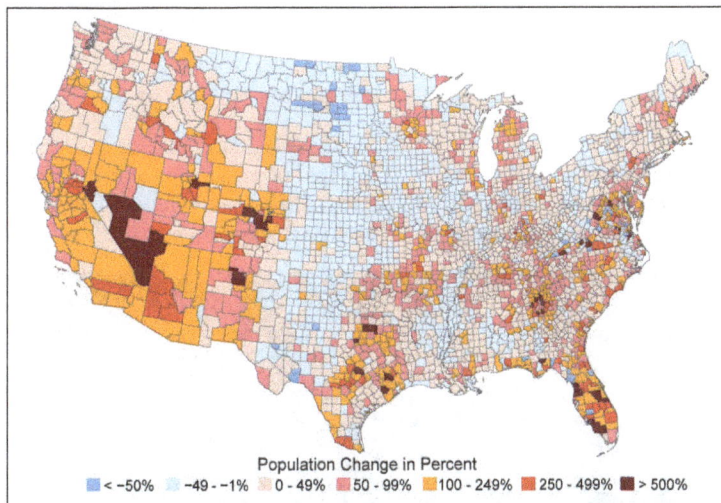

The percentage change in population across the United States from 1970 to 2008. In recent decades, population has grown rapidly in coastal areas and in the southern and western regions of the United States.

Impacts on Vulnerability and Equity

Projected climate change will affect certain groups of people more than others, depending on where they live and their ability to cope with different climate hazards. In some cases, the impacts of climate change are expected to worsen existing vulnerabilities.

Geographic Location

Where people live influences their vulnerability to climate change.

- Over the past four decades, population has grown rapidly in coastal areas and in the southern and western regions of the United States. These areas are most sensitive to coastal storms, drought, air pollution, and heat waves.

- Populations in the Mountain West will likely face water shortages and increased wildfires in the future.

- Arctic residents will likely experience problems caused by thawing permafrost and reduced sea ice.

- Along the coasts and across the western United States, both increasing population and changes in climate place growing demands on transportation, water, and energy infrastructure.

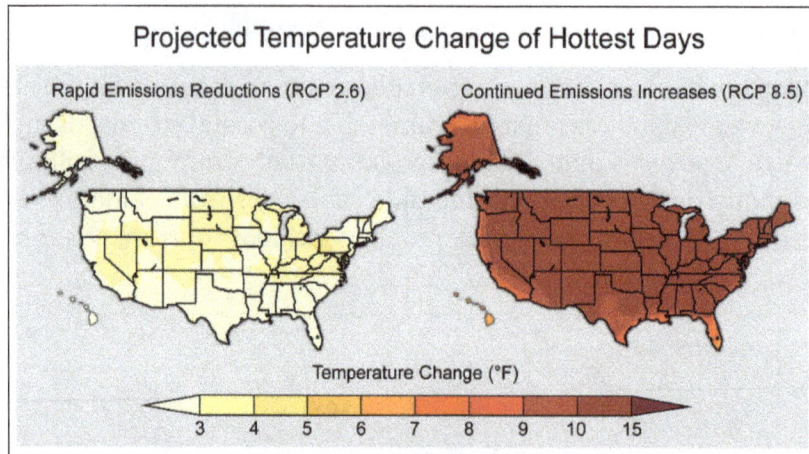

The average temperature on the hottest days (i.e., those that occur only once in 20 years) are projected to increase by the end of the century relative to 1986-2005. Those days will be 10 °F to 15 °F hotter under the "continued emissions increases" scenario by 2100.

Thousands of New Orleans evacuees relocated to the Houston Astrodome after Hurricane Katrina.

Ability to Cope

Different groups have different abilities to cope with climate change impacts.

- People who live in poverty may have a difficult time coping with changes. These people have limited financial resources to cope with heat, relocate or evacuate, or respond to increases in the cost of food.

- Older adults may be among the least able to cope with impacts of climate change.

Elderly people are particularly prone to heat stress.

- Older residents make up a larger share of the population in warmer areas of the United States. These areas will likely experience higher temperatures, tropical storms, or extended droughts in the future. The share of the U.S. population composed of adults over age 65 is also projected to grow from 13% in 2010 to 20% by 2050.

- Young children are another sensitive age group, since their immune system and other bodily systems are still developing and they rely on others to care for them in disaster situations.

Indigenous Peoples

Indigenous communities and tribes are diverse and span the United States. While each community and tribe is unique, many share characteristics that can affect their ability to prepare for, respond to, and cope with the impacts of climate change. These include:

- Living in rural areas or places most affected by climate change (like communities along the coast).

- Relying on surrounding environment and natural resources for food, cultural practices, and income.

- Coping with higher levels of existing health risks when compared to other groups.

- Having high rates of uninsured individuals, who have difficulty accessing quality health care.

- Living in isolated or low income communities.

Climate change can impact the health and well-being of indigenous tribes in many ways. Climate change will make it harder for tribes to access safe and nutritious food, including traditional foods important to many tribes' cultural practices. Many tribes already lack access to safe drinking water and wastewater treatment in their communities. Climate change is expected to increase health risks associated with water quality problems like contamination and may reduce availability of water, particularly during droughts.

By affecting the environment and natural resources of tribal communities, climate change also threatens the cultural identities of Indigenous people. As plants and animals used in traditional practices or sacred ceremonies become less available, tribal culture and ways of life can be greatly affected.

Urban Populations

City residents and urban infrastructure have distinct sensitivities to climate change impacts. For example, heat waves may be amplified in cities because cities absorb more heat during the day than suburban and rural areas.

Cities are more densely populated than suburban or rural areas. In fact, about 80% of the U.S. population lives in urban areas. As a result, increases in heat waves, drought, or violent storms in cities would affect a larger number of people than in suburban or rural areas. Higher temperatures and more extreme events will likely affect the cost of energy air and water quality, and human comfort and health in cities.

City dwellers may also be particularly susceptible to vulnerabilities in aging infrastructure. This includes drainage and sewer systems, flood and storm protection assets, transportation systems, and power supply during periods of peak demand, which typically occur during summer heat waves.

Impacts on Economic Activities and Services

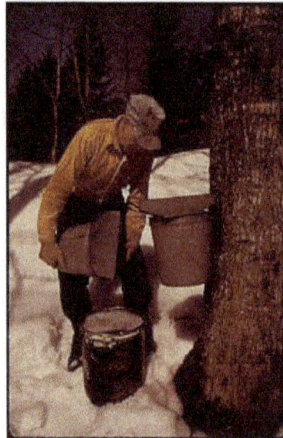

Sugarmaker harvesting maple syrup near Randolph Center, VT.

Certain areas of the United States benefit from being located close to natural resources that support the local economy. Climate change could threaten these resources, as well as the goods and services they produce and the jobs and livelihoods of those who depend upon them. For example, climate change will likely affect farming communities, tourism and recreation, and the insurance industry.

- Communities that developed around the production of different agricultural crops, such as corn, wheat, or cotton, depend on the climate to support their way of life. Climate change will likely cause the ideal climate for these crops to shift northward. Combined with decreasing rural populations, as in the Great Plains, a changing climate may fundamentally change many of these communities. Certain agricultural products, such as maple syrup and cranberries in the Northeast and grapes for wine in California, may decline dramatically in the U.S. These crops would then have to be imported.

- Climate change will also likely affect tourism and recreational activities. A warming climate and changes in precipitation patterns will likely decrease the number of days when recreational snow activities such as skiing and snowmobiling can take place. In the Southwest and Mountain West, an increasing number of wildfires could affect hiking and recreation in parks. Beaches could suffer erosion due to sea level rise and storm surge. Changes in the migration patterns of fish and animals would affect fishing and hunting. Communities that support themselves through these recreational activities would feel economic impacts as tourism patterns begin to change.

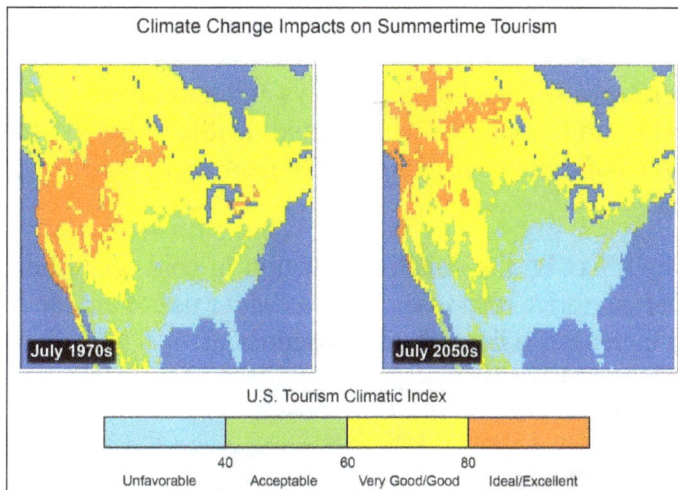

Summertime recreation and tourism is projected to become unfavorable across larger portions of the U.S. by 2050 due to increased July temperatures and humidity.

- Climate change may make it harder and more expensive for many people to insure their homes, businesses, or other valuable assets in risk-prone areas, or preclude them from insurance altogether. Insurance is one of the primary mechanisms used to protect people and communities against weather-related disasters. We rely on insurance to protect investments in real estate, agriculture, transportation, and utility infrastructure by distributing costs across society and build resilience. Climate change is projected to increase the frequency and intensity of extreme weather events, such as heat waves, droughts, and floods. These changes are likely to increase losses to property and crops, and cause costly disruptions to society. Escalating losses have already affected the availability and affordability of insurance in vulnerable areas.

Impacts of Climate Change on Human Health

The impacts of climate change include warming temperatures, changes in precipitation, increases in the frequency or intensity of some extreme weather events, and rising sea levels. These impacts threaten our health by affecting the food we eat, the water we drink, the air we breathe, and the weather we experience.

The severity of these health risks will depend on the ability of public health and safety systems to address or prepare for these changing threats, as well as factors such as an individual's behavior, age, gender, and economic status. Impacts will vary based on a where a person lives, how sensitive they are to health threats, how much they are exposed to climate change impacts, and how well they and their community are able to adapt to change.

People in developing countries may be the most vulnerable to health risks globally, but climate change poses significant threats to health even in wealthy nations such as the United States. Certain populations, such as children, pregnant women, older adults, and people with low incomes, face increased risks.

Temperature-related Impacts

Warmer average temperatures will lead to hotter days and more frequent and longer heat waves. These changes will lead to an increase in heat-related deaths in the United States—reaching as much as thousands to tens of thousands of additional deaths each year by the end of the century during summer months.

These deaths will not be offset by the smaller reduction in cold-related deaths projected in the winter months. However, adaptive responses, such as wider use of air conditioning, are expected to reduce the projected increases in death from extreme heat.

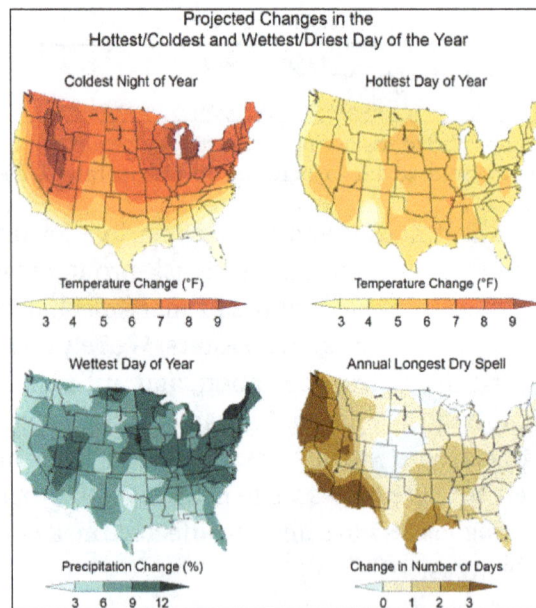

Projected changes in several climate variables for 2046-2065 with respect to the 1981-2000 average for the RCP6.0 scenario. These include the coldest night of the year (top left) and the hottest day of the year (top right). By the middle of this century, the coldest night of the year is projected to warm by 6°F to 10°F over most of the country, with slightly smaller changes in the south. The warmest day of the year is projected to be 4°F to 6°F warmer in most areas. Also shown are projections of the wettest day of the year (bottom left) and the annual longest consecutive dry day spell (bottom right). Extreme precipitation is projected to increase, with an average change of 5% to 15% in the precipitation falling on the wettest day of the year. The length of the annual longest dry spell is projected to increase in most areas, but these changes are small: less than two days in most areas.

Exposure to extreme heat can lead to heat stroke and dehydration, as well as cardiovascular, respiratory, and cerebrovascular disease. Excessive heat is more likely to affect populations in northern latitudes where people are less prepared to cope with excessive temperatures. Certain types of populations are more vulnerable than others: for example, outdoor workers, student athletes, and homeless people tend to be more exposed to extreme heat because they spend more time outdoors. Low-income households and older adults may lack access to air conditioning which also increases exposure to extreme heat. Additionally, young children, pregnant women, older adults, and people with certain medical conditions are less able to regulate their body temperature and can therefore be more vulnerable to extreme heat.

Urban areas are typically warmer than their rural surroundings. Large metropolitan areas such as St. Louis, Philadelphia, Chicago, and Cincinnati have seen notable increases in death rates during heat waves. Climate change is projected to increase the vulnerability of urban populations to heat-related health impacts in the future. Heat waves are also often accompanied by periods of stagnant air, leading to increases in air pollution and associated health effects.

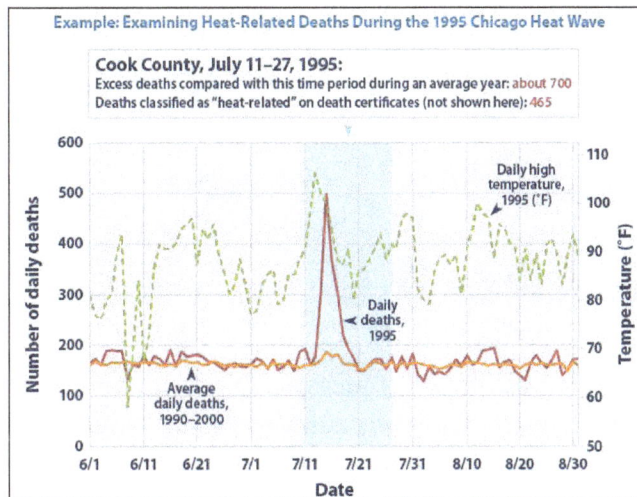

This figure shows the relationship between high temperatures and deaths observed during the 1995 Chicago heat wave. The large spike in deaths in mid-July (red line) is much higher than the average number of deaths during that time of year (orange line), as well as the death rate before and after the heat wave.

Air Quality Impacts

Changes in the climate affect the air we breathe both indoors and outdoors. Warmer temperatures and shifting weather patterns can worsen air quality, which can lead to asthma attacks and other respiratory and cardiovascular health effects. Wildfires, which are expected to continue to increase in number and severity as the climate changes, create smoke and other unhealthy air pollutants. Rising carbon dioxide levels and warmer temperatures also affect airborne allergens, such as ragweed pollen.

Despite significant improvements in U.S. air quality since the 1970s, as of 2014 about 57 million Americans lived in counties that did not meet national air quality standards. Climate change may make it even harder for states to meet these standards in the future, exposing more people to unhealthy air.

Increases in Ozone

Scientists project that warmer temperatures from climate change will increase the frequency of days with unhealthy levels of ground-level ozone, a harmful air pollutant, and a component in smog.

- People exposed to higher levels of ground-level ozone are at greater risk of dying prematurely or being admitted to the hospital for respiratory problems.

- Ground-level ozone can damage lung tissue, reduce lung function, and inflame airways. This can aggravate asthma or other lung diseases. Children, older adults, outdoor workers, and those with asthma and other chronic lung diseases are particularly at risk.

Smog decreases visibility and can be harmful to human health.

- Because warm, stagnant air tends to increase the formation of ozone, climate change is likely to increase levels of ground-level ozone in already-polluted areas of the United States and increase the number of days with poor air quality.

- The higher concentrations of ozone due to climate change may result in tens to thousands of additional ozone-related illnesses and premature deaths per year by 2030 in the United States, assuming no change in projected air quality policies.

Changes in Particulate Matter

Particulate matter is the term for a category of extremely small particles and liquid droplets suspended in the atmosphere. Fine particles include those smaller than 2.5 micrometers (about one ten-thousandth of an inch). Some particulate matter such as dust, wildfire smoke, and sea spray occur naturally, while some is created by human activities such as the burning of fossil fuels to produce energy. These particles may be emitted directly or may be formed in the atmosphere from chemical reactions of gases such as sulfur dioxide, nitrogen dioxide, and volatile organic compounds.

- Inhaling fine particles can lead to a broad range of adverse health effects, including lung cancer, chronic obstructive pulmonary disease (COPD), and cardiovascular disease.

- Climate change is expected to increase the number and severity of wildfires. Particulate matter from wildfire smoke can often be carried very long distances by the wind, affecting people who live far from the source of this air pollutant.

- Older adults are particularly sensitive to short-term particle exposure, with a higher risk of hospitalization and death. Outdoor workers like firefighters can also have high exposure.

Due to the complex factors that influence atmospheric levels of fine particulate matter, scientists do not yet know whether climate change will increase or decrease particulate matter concentrations across the United States. Particulate matter can be removed from the air by rainfall, and precipitation is expected to increase in quantity though not necessarily frequency. Climate-related changes in stagnant air episodes, wind patterns, emissions from vegetation and the chemistry of atmospheric pollutants will also affect particulate matter levels.

Changes in Allergens and Asthma Triggers

Allergic illnesses, including hay fever, affect about one-third of the U.S. population, and more than 34 million Americans have been diagnosed with asthma. Climate change may affect allergies and respiratory health. The spring pollen season is already occurring earlier in the United States for certain types of plants, and the length of the season has increased for some plants with highly allergenic pollen such as ragweed. In addition to lengthening the ragweed pollen season, rising carbon dioxide concentrations and temperatures may also lead to earlier flowering, more flowers, and increased pollen levels in ragweed.

Impacts from Extreme Weather Events

Increases in the frequency or severity of some extreme weather events, such as extreme precipitation, flooding, droughts, and storms, threaten the health of people during and after the event. The people most at risk include young children, older adults, people with disabilities or medical conditions, and the poor. Extreme events can affect human health in a number of ways by:

Hurricane Katrina was one of the most devastating hurricanes in the
United States, responsible for an estimated 971 to 1,300 deaths.

- Reducing the availability of safe food and drinking water.

- Damaging roads and bridges, disrupting access to hospitals and pharmacies.

- Interrupting communication, utility, and health care services.

- Contributing to carbon monoxide poisoning from improper use of portable electric generators during and after storms.

- Increasing stomach and intestinal illness, particularly following power outages.

- Creating or worsening mental health impacts such as depression and post-traumatic stress disorder (PTSD).

In addition, emergency evacuations pose health risks to older adults, especially those with limited mobility who cannot use elevators during power outages. Evacuations may be complicated by the need for concurrent transfer of medical records, medications, and medical equipment. Some individuals with disabilities may also be disproportionally affected if they are unable to access evacuation routes, have difficulty in understanding or receiving warnings of impending danger, or have limited ability to communicate their needs.

Vectorborne Diseases

Vectorborne diseases are illnesses that are transmitted by disease vectors, which include mosquitoes, ticks, and fleas. These vectors can carry infectious pathogens, such as viruses, bacteria, and protozoa, from animals to humans. Changes in temperature, precipitation, and extreme events increases the geographic range of diseases spread by vectors and can lead to illnesses occurring earlier in the year.

- The geographic range of ticks that carry Lyme disease is limited by temperature. As air temperatures rise, ticks are likely to become active earlier in the season, and their range is likely to continue to expand northward. Typical symptoms of Lyme disease include fever, headache, fatigue, and a characteristic skin rash.

- Mosquitoes thrive in certain climate conditions and can spread diseases like West Nile virus. Extreme temperatures—too cold, hot, wet, or dry—influence the location and number of mosquitoes that transmit West Nile virus. More than three million people were estimated to be infected with West Nile virus in the United States from 1999 to 2010.

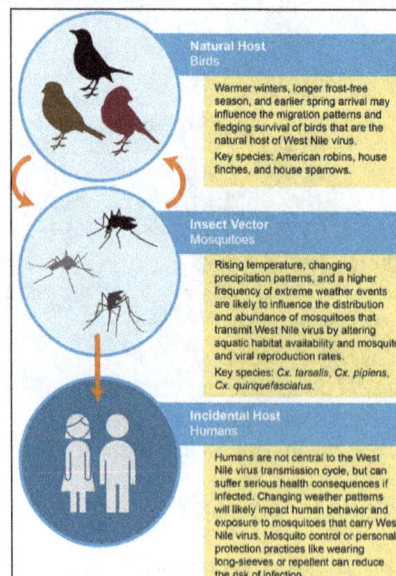

The spread of climate-sensitive diseases will depend on both climate and non-climate factors such as land use, socioeconomic and cultural conditions, pest control, access to health care, and human responses to disease risk. The United States has public health infrastructure and programs to monitor, manage, and prevent the spread of many diseases. The risks for climate-sensitive diseases can be much higher in poorer countries that have less capacity to prevent and treat illness.

West Nile virus is maintained in transmission cycles between birds (the natural hosts of the virus) and mosquitoes. Human infections can occur from a bite of a mosquito that has previously bitten an infected bird. Warmer winters, longer frost-free season, and earlier spring arrival may influence the migration patterns and fledgling survival of birds that are the natural host of West Nile virus. In addition, rising temperature, changing precipitation patterns, and a higher frequency of extreme weather events are likely to influence the distribution and abundance of mosquitoes that transmit West Nile virus.

Water-related Illnesses

People can become ill if exposed to contaminated drinking or recreational water. Climate change increases the risk of illness through increasing temperature, more frequent heavy rains and runoff, and the effects of storms. Health impacts may include gastrointestinal illness like diarrhea, effects on the body's nervous and respiratory systems, or liver and kidney damage.

- Climate impacts can affect exposure to waterborne pathogens (bacteria, viruses, and parasites such as *Cryptosporidium* and *Giardia*); toxins produced by harmful algal and cyanobacterial blooms in the water; and chemicals that end up in water from human activities.

- Changing water temperatures mean that waterborne *Vibrio* bacteria and harmful algal toxins will be present in the water or in seafood at different times of the year, or in places where they were not previously threats.

- Runoff and flooding resulting from increases in extreme precipitation, hurricane rainfall, and storm surge will increasingly contaminate water bodies used for recreation (such as lakes and beaches), shellfish harvesting waters, and sources of drinking water.

- Extreme weather events and storm surges can damage or exceed the capacity of water infrastructure (such as drinking water or wastewater treatment plants), increasing the risk that people will be exposed to contaminants.

Water resource, public health, and environmental agencies in the United States provide many public health safeguards to reduce risk of exposure and illness even if water becomes contaminated. These include water quality monitoring, drinking water treatment standards and practices, beach closures, and issuing advisories for boiling drinking water and harvesting shellfish.

Food Safety and Nutrition

Climate change and the direct impacts of higher concentrations of carbon dioxide in the atmosphere are expected to affect food safety and nutrition. Extreme weather events can also disrupt or slow the distribution of food.

- Higher air temperatures can increase cases of *Salmonella* and other bacteria-related food poisoning because bacteria grow more rapidly in warm environments. These diseases can cause gastrointestinal distress and, in severe cases, death. Practices to safeguard food can help avoid these illnesses even as the climate changes.

- Climate change will have a variety of impacts that may increase the risk of exposure to chemical contaminants in food. For example, higher sea surface temperatures will lead to higher mercury concentrations in seafood, and increases in extreme weather events will introduce contaminants into the food chain through stormwater runoff.

- Higher concentrations of carbon dioxide in the air can act as a "fertilizer" for some plants, but lowers the levels of protein and essential minerals in crops such as wheat, rice, and potatoes, making these foods less nutritious.

- Extreme events, such as flooding and drought, create challenges for food distribution if roads and waterways are damaged or made inaccessible.

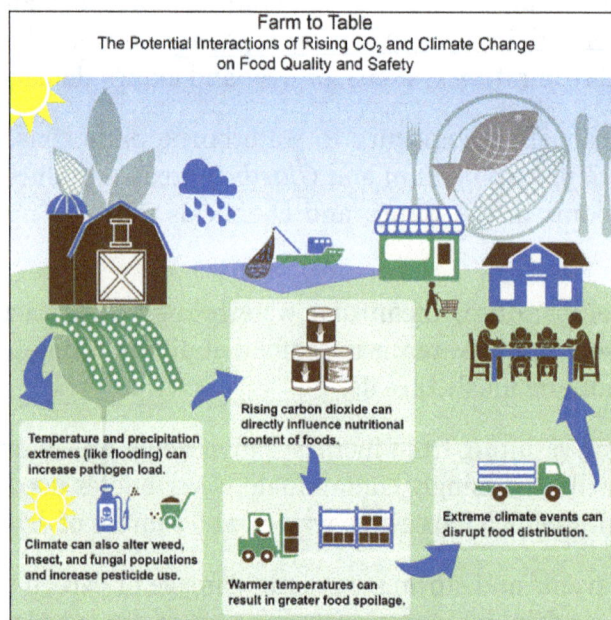

The food system involves a network of interactions with our physical and biological environments as food moves from production to consumption, or from "farm to table." Rising CO_2 and climate change will affect the quality and distribution of food, with subsequent effects on food safety and nutrition.

Mental Health

Any changes in a person's physical health or surrounding environment can also have serious impacts on their mental health. In particular, experiencing an extreme weather event can cause stress and other mental health consequences, particularly when a person loses loved ones or their home.

- Individuals with mental illness are especially vulnerable to extreme heat; studies have found that having a pre-existing mental illness tripled the risk of death during heat waves. People taking medication for mental illness that makes it difficult to regulate their body temperature are particularly at risk.

- Even the perceived threat of climate change (for example from reading or watching news reports about climate change) can influence stress responses and mental health.

- Some groups of people are at higher risk for mental health impacts, such as children and older adults, pregnant and post-partum women, people with pre-existing mental illness, people with low incomes, and emergency workers.

Populations of Concern

Some groups of people are more vulnerable than others to health risks from climate change. Three factors contribute to vulnerability: *sensitivity*, which refers to the degree to which people or groups are affected by a stressor such as higher temperatures; *exposure*, which refers to physical contact between a person and a stressor; and *adaptive capacity*, which refers to an ability to adjust to or avoid potential hazards. For example, while older adults are sensitive to extreme heat, an older person living in an air-conditioned apartment won't be exposed as long as she stays indoors, and as long as she can afford to pay for the electricity to run the air conditioner. Her ability take these actions is a measure of her adaptive capacity.

Some populations are especially vulnerable to climate health risks due to particular sensitivities, high likelihood of exposure, low adaptive capacity, or combinations of these factors.

- Communities of color (including Indigenous communities as well as specific racial and ethnic groups), low income, immigrants, and limited English proficiency face disproportionate vulnerabilities due to a wide variety of factors, such as higher risk of exposure, socioeconomic and educational factors that affect their adaptive capacity, and a higher prevalence of medical conditions that affect their sensitivity.

- Children are vulnerable to many health risks due to biological sensitivities and more opportunities for exposure (due to activities such as playing outdoors). Pregnant women are vulnerable to heat waves and other extreme events, like flooding.

- Older adults are vulnerable to many of the impacts of climate change. They may have greater sensitivity to heat and contaminants, a higher prevalence of disability or preexisting medical conditions, or limited financial resources that make it difficult to adapt to impacts.

- Occupational groups, such as outdoor workers, paramedics, firefighters, and transportation workers, as well as workers in hot indoor work environments, will be especially vulnerable to extreme heat and exposure to vectorborne diseases.

- People with disabilities can be very vulnerable during extreme weather events, unless communities ensure that their emergency response plans specifically accommodate them.

- People with chronic medical conditions are typically vulnerable to extreme heat, especially if they are taking medications that make it difficult to regulate body temperature. Power outages can be particularly threatening for people reliant on certain medical equipment.

Other Health Impacts

Other linkages exist between climate change and human health. For example, changes in temperature and precipitation, as well as droughts and floods, will affect agricultural yields and production. In some regions of the world, these impacts may compromise food security and threaten human health through malnutrition, the spread of infectious diseases, and food poisoning. The worst of these effects are projected to occur in developing countries, among vulnerable populations. Declines in human health in other countries can affect the United States through trade, migration, and immigration and has implications for national security.

Although the impacts of climate change have the potential to affect human health in the United States and around the world, there is a lot we can do to prepare for and adapt to these changes—such as establishing early warning systems for heat waves and other extreme events, taking steps to reduce vulnerabilities among populations of concern, raising awareness among healthcare professionals, and ensuring that infrastructure is built to accommodate anticipated future changes in climate. Understanding the threats that climate change poses to human health is the first step in working together to lower risks and be prepared.

Air Pollution and Climate Change

Air pollution changes our planet's climate, but not all types of air pollution have the same effect. There are many different types of air pollution. Some types cause global warming to speed up. Others cause global warming to slow down by creating a temporary cooling effect for a few days or weeks. Beyond that, we are emitting such a high level of pollutants that they are causing serious global environmental problems: climate change and ozone depletion. The human race has become capable of affecting the atmosphere that encircles the Earth, and the very planet itself.

Both climate change and air pollution are worsened by the burning of fuel, increasing the CO_2 emissions which cause global warming. Meanwhile, the generation of other pollutants, such as nitrogen oxides (NO and NO_2), sulfur oxides (SO_2 and SO_3) and particulate matter, is the main reason the air is contaminated.

Global Health Hazard

While those effects emerge from long-term exposure, air pollution can also cause short-term problems such as sneezing and coughing, eye irritation, headaches, and dizziness. Particulate matter smaller than 10 micrometers (classified as PM_{10} and the even smaller $PM_{2.5}$) pose higher health risks because they can be breathed deeply into the lungs and may cross into the bloodstream.

Air pollutants cause less-direct health effects when they contribute to climate change. Heat waves, extreme weather, food supply disruptions, and other effects related to increased greenhouse gases can have negative impacts on human health. The U.S. Fourth National Climate Assessment released in 2018 noted, for example, that a changing climate "could expose more people in North America to ticks that carry Lyme disease and mosquitoes that transmit viruses such as West Nile, chikungunya, dengue, and Zika".

Environmental Impacts

Though many living things emit carbon dioxide when they breathe, the gas is widely considered to be a pollutant when associated with cars, planes, power plants, and other human activities that involve the burning of fossil fuels such as gasoline and natural gas. That's because carbon dioxide is the most common of the greenhouse gases, which trap heat in the atmosphere and contribute to climate change. Humans have pumped enough carbon dioxide into the atmosphere over the past 150 years to raise its levels higher than they have been for hundreds of thousands of years.

Other greenhouse gases include methane —which comes from such sources as landfills, the natural gas industry, and gas emitted by livestock—and chlorofluorocarbons (CFCs), which were used in refrigerants and aerosol propellants until they were banned in the late 1980s because of their deteriorating effect on Earth's ozone layer.

Greenhouse gases are a key factor in the Earth's changing climate.

Another pollutant associated with climate change is sulfur dioxide, a component of smog. Sulfur dioxide and closely related chemicals are known primarily as a cause of acid rain. But they also reflect light when released in the atmosphere, which keeps sunlight out and creates a cooling effect. Volcanic eruptions can spew massive amounts of sulfur dioxide into the atmosphere, sometimes causing cooling that lasts for years. In fact, volcanoes used to be the main source of atmospheric sulfur dioxide; today, people are.

Airborne particles, depending on their chemical makeup, can also have direct effects separate from climate change. They can change or deplete nutrients in soil and waterways, harm forests and crops, and damage cultural icons such as monuments and statues.

Acid Rain

Acid rain, or acid deposition, is a broad term that includes any form of precipitation with acidic components, such as sulfuric or nitric acid that fall to the ground from the atmosphere in wet or dry forms. This can include rain, snow, fog, hail or even dust that is acidic.

Causes of Acid Rain

Acid rain results when sulfur dioxide (SO_2) and nitrogen oxides (NO_x) are emitted into the atmosphere and transported by wind and air currents. The SO_2 and NO_X react with water, oxygen and other chemicals to form sulfuric and nitric acids. These then mix with water and other materials before falling to the ground.

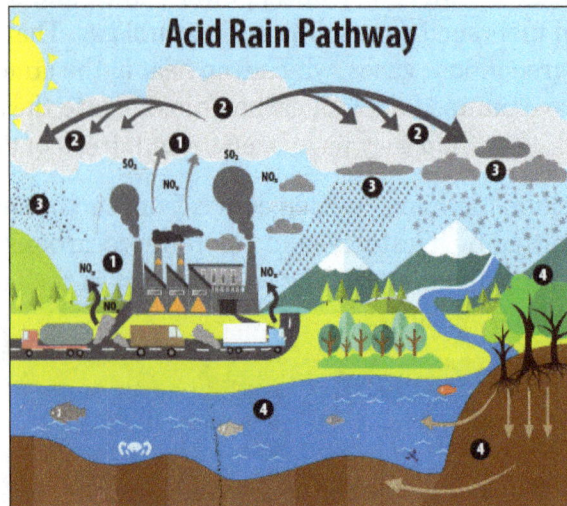

(1) Emissions io SO2 and NOx are released into the air, where (2) the pollutants are transformed into acid particles that may be transported long distances. (3) These acid particles then fall to the earth as wet and dry deposition (dust, rain, etc.) and (4) may cause harmful effects on soil, forests, streams and lakes.

While a small portion of the SO_2 and NO_X that cause acid rain is from natural sources such as volcanoes, most of it comes from the burning of fossil fuels. The major sources of SO_2 and NO_X in the atmosphere are:

- Burning of fossil fuels to generate electricity. Two thirds of SO_2 and one fourth of NO_X in the atmosphere come from electric power generators.

- Vehicles and heavy equipment.

- Manufacturing, oil refineries and other industries.

Winds can blow SO_2 and NO_X over long distances and across borders making acid rain a problem for everyone and not just those who live close to these sources.

Forms of Acid Deposition

Wet Deposition

Wet deposition is what we most commonly think of as acid rain. The sulfuric and nitric acids formed in the atmosphere fall to the ground mixed with rain, snow, fog, or hail.

Dry Deposition

Acidic particles and gases can also deposit from the atmosphere in the absence of moisture as dry deposition. The acidic particles and gases may deposit to surfaces (water bodies, vegetation, buildings) quickly or may react during atmospheric transport to form larger particles that can be harmful to human health. When the accumulated acids are washed off a surface by the next rain, this acidic water flows over and through the ground, and can harm plants and wildlife, such as insects and fish.

The amount of acidity in the atmosphere that deposits to earth through dry deposition depends on the amount of rainfall an area receives. For example, in desert areas the ratio of dry to wet deposition is higher than an area that receives several inches of rain each year.

Measuring Acid Rain

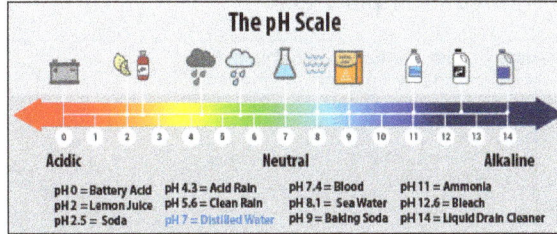

Acidity and alkalinity are measured using a pH scale for which 7.0 is neutral. The lower a substance's pH (less than 7), the more acidic it is; the higher a substance's pH (greater than 7), the more alkaline it is. Normal rain has a pH of about 5.6; it is slightly acidic because carbon dioxide (CO_2) dissolves into it forming weak carbonic acid. Acid rain usually has a pH between 4.2 and 4.4.

Global Warming

Global warming is the phenomenon of increasing average air temperatures near the surface of Earth over the past one to two centuries. Climate scientists have since the mid-20th century gathered detailed observations of various weather phenomena (such as temperatures, precipitation, and storms) and of related influences on climate (such as ocean currents and the atmosphere's chemical composition). These data indicate that Earth's climate has changed over almost every conceivable timescale since the beginning of geologic time and that the influence of human activities since at least the beginning of the Industrial Revolution has been deeply woven into the very fabric of climate change.

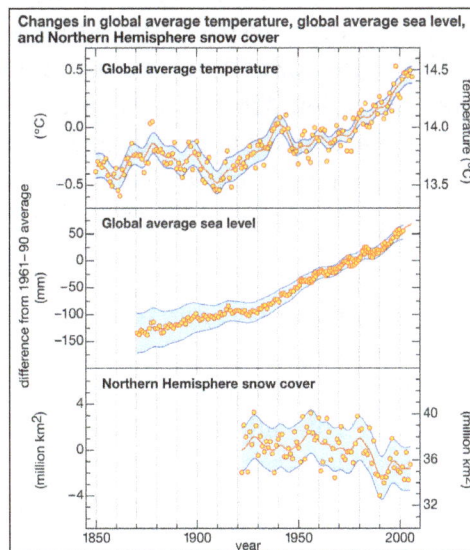

During the second half of the 20th century and early part of the 21st century, global average surface temperature increased and sea level rose. Over the same period, the amount of snow cover in the Northern Hemisphere decreased.

Giving voice to a growing conviction of most of the scientific community, the Intergovernmental Panel on Climate Change (IPCC) was formed in 1988 by the World Meteorological Organization (WMO) and the United Nations Environment Program (UNEP). In 2013 the IPCC reported that the interval between 1880 and 2012 saw an increase in global average surface temperature of approximately 0.9 °C (1.5 °F). The increase is closer to 1.1 °C (2.0 °F) when measured relative to the preindustrial (i.e., 1750–1800) mean temperature.

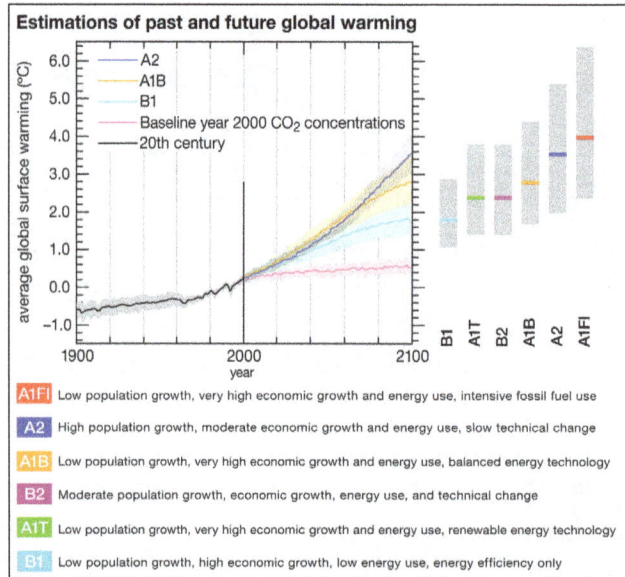

Global Warming Scenarios Graph of the predicted increase in Earth's average surface temperature according to a series of climate change scenarios that assume different levels of economic development, population growth, and fossil fuel use. The assumptions made by each scenario are given at the bottom of the graph.

A special report produced by the IPCC in 2018 honed this estimate further, noting that human beings and human activities have been responsible for a worldwide average temperature increase of between 0.8 and 1.2 °C (1.4 and 2.2 °F) of global warming since preindustrial times, and most of the warming observed over the second half of the 20th century could be attributed to human activities. It predicted that the global mean surface temperature would increase between 3 and 4 °C (5.4 and 7.2 °F) by 2100 relative to the 1986–2005 average should carbon emissions continue at their current rate. The predicted rise in temperature was based on a range of possible scenarios that accounted for future greenhouse gas emissions and mitigation (severity reduction) measures and on uncertainties in the model projections. Some of the main uncertainties include the precise role of feedback processes and the impacts of industrial pollutants known as aerosols, which may offset some warming.

Many climate scientists agree that significant societal, economic, and ecological damage would result if global average temperatures rose by more than 2 °C (3.6 °F) in such a short time. Such damage would include increased extinction of many plant and animal species, shifts in patterns of agriculture, and rising sea levels. By 2015 all but a few national governments had begun the process of instituting carbon reduction plans as part of the Paris Agreement, a treaty designed to help countries keep global warming to 1.5 °C (2.7 °F) above preindustrial levels in order to avoid the worst of the predicted effects.

The scenarios referred to above depend mainly on future concentrations of certain trace gases, called greenhouse gases, that have been injected into the lower atmosphere in increasing amounts through the burning of fossil fuels for industry, transportation, and residential uses. Modern global warming is the result of an increase in magnitude of the so-called greenhouse effect, a warming of Earth's surface and lower atmosphere caused by the presence of water vapour, carbon dioxide, methane, nitrous oxides, and other greenhouse gases. In 2014 the IPCC reported that concentrations of carbon dioxide, methane, and nitrous oxides in the atmosphere surpassed those found in ice cores dating back 800,000 years.

Causes of Global Warming

Greenhouse Effect

The average surface temperature of Earth is maintained by a balance of various forms of solar and terrestrial radiation. Solar radiation is often called "shortwave" radiation because the frequencies of the radiation are relatively high and the wavelengths relatively short—close to the visible portion of the electromagnetic spectrum. Terrestrial radiation, on the other hand, is often called "longwave" radiation because the frequencies are relatively low and the wavelengths relatively long—somewhere in the infrared part of the spectrum. Downward-moving solar energy is typically measured in watts per square metre. The energy of the total incoming solar radiation at the top of Earth's atmosphere (the so-called "solar constant") amounts roughly to 1,366 watts per square metre annually. Adjusting for the fact that only one-half of the planet's surface receives solar radiation at any given time, the average surface insolation is 342 watts per square metre annually.

The amount of solar radiation absorbed by Earth's surface is only a small fraction of the total solar radiation entering the atmosphere. For every 100 units of incoming solar radiation, roughly 30 units are reflected back to space by either clouds, the atmosphere, or reflective regions of Earth's surface. This reflective capacity is referred to as Earth's planetary albedo, and it need not remain fixed over time, since the spatial extent and distribution of reflective formations, such as clouds and ice cover, can change. The 70 units of solar radiation that are not reflected may be absorbed by the atmosphere, clouds, or the surface. In the absence of further complications, in order to maintain thermodynamic equilibrium, Earth's surface and atmosphere must radiate these same 70 units back to space. Earth's surface temperature (and that of the lower layer of the atmosphere essentially in contact with the surface) is tied to the magnitude of this emission of outgoing radiation according to the Stefan-Boltzmann law.

Earth's energy budget is further complicated by the greenhouse effect. Trace gases with certain chemical properties—the so-called greenhouse gases, mainly carbon dioxide (CO_2), methane (CH_4), and nitrous oxide (N_2O)—absorb some of the infrared radiation produced by Earth's surface. Because of this absorption, some fraction of the original 70 units does not directly escape to space. Because greenhouse gases emit the same amount of radiation they absorb and because this radiation is emitted equally in all directions (that is, as much downward as upward), the net effect of absorption by greenhouse gases is to increase the total amount of radiation emitted downward toward Earth's surface and lower atmosphere. To maintain equilibrium, Earth's surface and lower atmosphere must emit more radiation than the original 70 units. Consequently, the surface temperature must be higher. This process is not quite the same as that which governs a true greenhouse, but the end effect is similar. The presence of greenhouse gases in the atmosphere leads to a warming of the surface and lower part of the atmosphere (and a cooling higher up in the atmosphere) relative to what would be expected in the absence of greenhouse gases.

It is essential to distinguish the "natural," or background, greenhouse effect from the "enhanced" greenhouse effect associated with human activity. The natural greenhouse effect is associated with surface warming properties of natural constituents of Earth's atmosphere, especially water vapour, carbon dioxide, and methane. The existence of this effect is accepted by all scientists. Indeed, in its absence, Earth's average temperature would be approximately 33 °C (59 °F) colder than today, and Earth would be a frozen and likely uninhabitable planet. What has been subject to controversy is the so-called enhanced greenhouse effect, which is associated with increased concentrations of greenhouse gases caused by human activity. In particular, the burning of fossil fuels raises the concentrations of the major greenhouse gases in the atmosphere, and these higher concentrations have the potential to warm the atmosphere by several degrees.

Radiative Forcing

It is apparent that the temperature of Earth's surface and lower atmosphere may be modified in three ways: (1) through a net increase in the solar radiation entering at the top of Earth's atmosphere, (2) through a change in the fraction of the radiation reaching the surface, and (3) through a change in the concentration of greenhouse gases in the atmosphere. In each case the changes can be thought of in terms of "radiative forcing." As defined by the IPCC, radiative forcing is a measure of the influence a given climatic factor has on the amount of downward-directed radiant energy impinging upon Earth's surface. Climatic factors are divided between those caused primarily by human activity (such as greenhouse gas emissions and aerosol emissions) and those caused by natural forces (such as solar irradiance); then, for each factor, so-called forcing values are calculated for the time period between 1750 and the present day. "Positive forcing" is exerted by climatic factors that contribute to the warming of Earth's surface, whereas "negative forcing" is exerted by factors that cool Earth's surface.

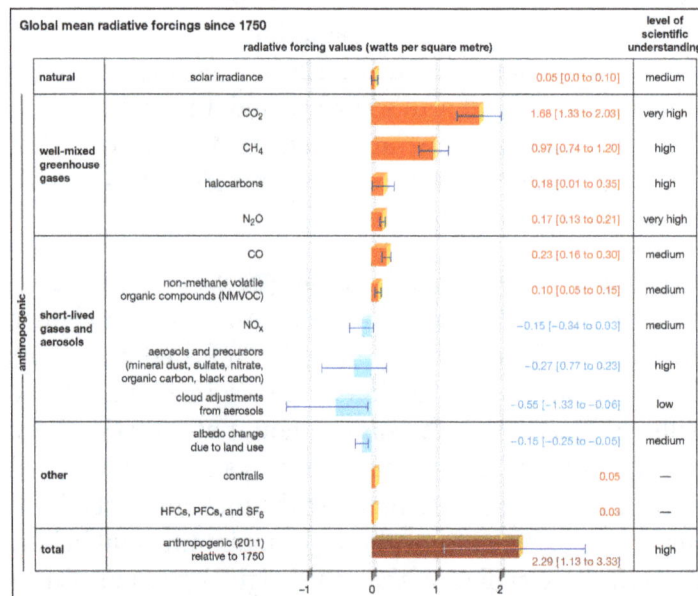

Global mean radiative forcings since 1750		radiative forcing values (watts per square metre)	level of scientific understanding
natural	solar irradiance	0.05 [0.0 to 0.10]	medium
well-mixed greenhouse gases	CO₂	1.68 [1.33 to 2.03]	very high
	CH₄	0.97 [0.74 to 1.20]	high
	halocarbons	0.18 [0.01 to 0.35]	high
	N₂O	0.17 [0.13 to 0.21]	very high
short-lived gases and aerosols	CO	0.23 [0.16 to 0.30]	medium
	non-methane volatile organic compounds (NMVOC)	0.10 [0.05 to 0.15]	medium
	NOₓ	−0.15 [−0.34 to 0.03]	medium
	aerosols and precursors (mineral dust, sulfate, nitrate, organic carbon, black carbon)	−0.27 [0.77 to 0.23]	high
	cloud adjustments from aerosols	−0.55 [−1.33 to −0.06]	low
other	albedo change due to land use	−0.15 [−0.25 to −0.05]	medium
	contrails	0.05	—
	HFCs, PFCs, and SF₆	0.03	—
total	anthropogenic (2011) relative to 1750	2.29 [1.13 to 3.33]	high

Since 1750 the concentration of carbon dioxide and other greenhouse gases has increased in Earth's atmosphere. As a result of these and other factors, Earth's atmosphere retains more heat than in the past.

On average, about 342 watts of solar radiation strike each square metre of Earth's surface per year, and this quantity can in turn be related to a rise or fall in Earth's surface temperature. Temperatures at the surface may also rise or fall through a change in the distribution of terrestrial radiation (that is, radiation emitted by Earth) within the atmosphere. In some cases, radiative forcing has a natural origin, such as during explosive eruptions from volcanoes where vented gases and ash block some portion of solar radiation from the surface. In other cases, radiative forcing has an anthropogenic, or exclusively human, origin. For example, anthropogenic increases in carbon dioxide, methane, and nitrous oxide are estimated to account for 2.3 watts per square metre of positive radiative forcing. When all values of positive and negative radiative forcing are taken together and all interactions between climatic factors are accounted for, the total net increase in surface radiation due to human activities since the beginning of the Industrial Revolution is 1.6 watts per square metre.

Influences of Human Activity on Climate

Petroleum refinery at Ras Tanura.

Human activity has influenced global surface temperatures by changing the radiative balance governing the Earth on various timescales and at varying spatial scales. The most profound and well-known anthropogenic influence is the elevation of concentrations of greenhouse gases in the atmosphere. Humans also influence climate by changing the concentrations of aerosols and ozone and by modifying the land cover of Earth's surface.

Greenhouse Gases

Factories that burn fossil fuels help to cause global warming.

As discussed before, greenhouse gases warm Earth's surface by increasing the net downward longwave radiation reaching the surface. The relationship between atmospheric concentration of greenhouse gases and the associated positive radiative forcing of the surface is different for each gas. A complicated relationship exists between the chemical properties of each greenhouse gas and the relative amount of longwave radiation that each can absorb.

Water Vapour

Water vapour is the most potent of the greenhouse gases in Earth's atmosphere, but its behaviour is fundamentally different from that of the other greenhouse gases. The primary role of water vapour is not as a direct agent of radiative forcing but rather as a climate feedback—that is, as a response within the climate system that influences the system's continued activity. This distinction arises from the fact that the amount of water vapour in the atmosphere cannot, in general, be directly modified by human behaviour but is instead set by air temperatures. The warmer the surface, the greater the evaporation rate of water from the surface. As a result, increased evaporation leads to a greater concentration of water vapour in the lower atmosphere capable of absorbing longwave radiation and emitting it downward.

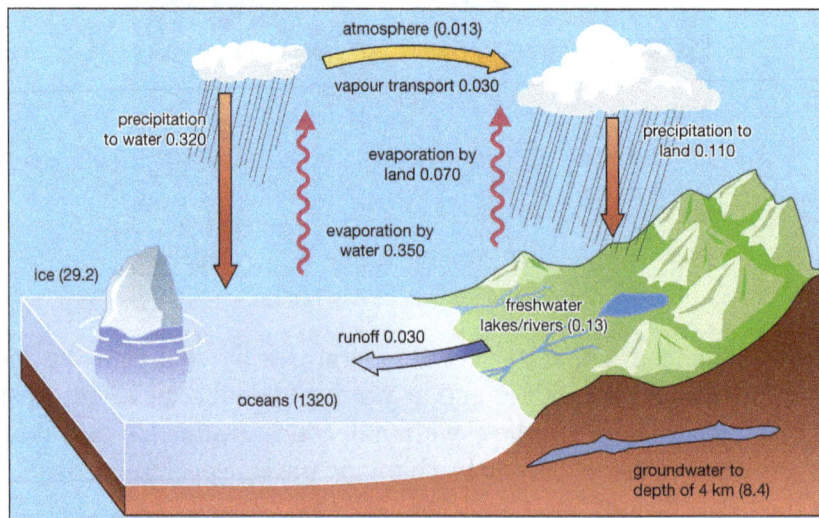

The present day surface hydrologic cycle, in which water is transferred from the oceans through the atmosphere to the continents and back to the oceans over and beneath the land surface. The values in parentheses following the various forms of water (e.g., ice) refer to volumes in millions of cubic kilometres; those following the processes (e.g., precipitation) refer to their fluxes in millions of cubic kilometres of water per year.

Carbon Dioxide

Of the greenhouse gases, carbon dioxide (CO_2) is the most significant. Natural sources of atmospheric CO_2 include outgassing from volcanoes, the combustion and natural decay of organic matter, and respiration by aerobic (oxygen-using) organisms. These sources are balanced, on average, by a set of physical, chemical, or biological processes, called "sinks," that tend to remove CO_2 from the atmosphere. Significant natural sinks include terrestrial vegetation, which takes up CO_2 during the process of photosynthesis.

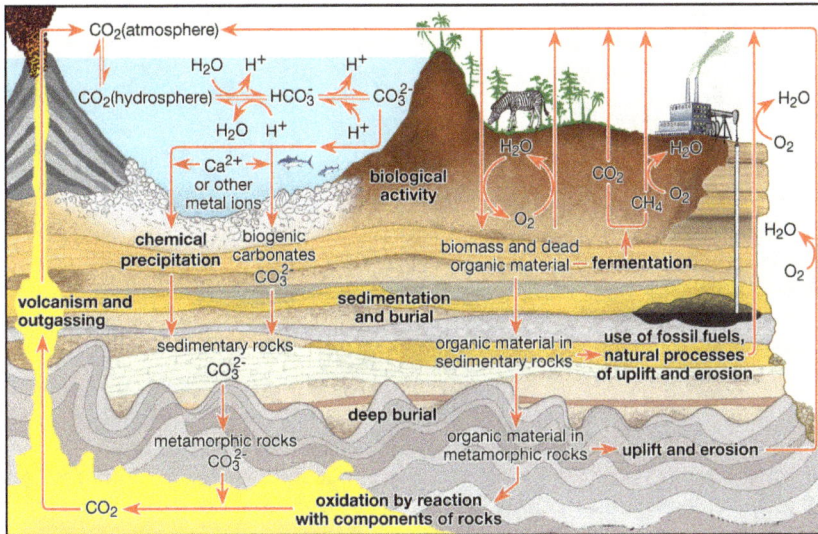

The carbon cycle.

Carbon is transported in various forms through the atmosphere, the hydrosphere, and geologic formations. One of the primary pathways for the exchange of carbon dioxide (CO_2) takes place between the atmosphere and the oceans; there a fraction of the CO_2 combines with water, forming carbonic acid (H_2CO_3) that subsequently loses hydrogen ions (H^+) to form bicarbonate (HCO_3^-) and carbonate (CO_3^{2-}) ions. Mollusk shells or mineral precipitates that form by the reaction of calcium or other metal ions with carbonate may become buried in geologic strata and eventually release CO_2 through volcanic outgassing. Carbon dioxide also exchanges through photosynthesis in plants and through respiration in animals. Dead and decaying organic matter may ferment and release CO_2 or methane (CH_4) or may be incorporated into sedimentary rock, where it is converted to fossil fuels. Burning of hydrocarbon fuels returns CO_2 and water (H_2O) to the atmosphere. The biological and anthropogenic pathways are much faster than the geochemical pathways and, consequently, have a greater impact on the composition and temperature of the atmosphere.

A number of oceanic processes also act as carbon sinks. One such process, called the "solubility pump," involves the descent of surface seawater containing dissolved CO_2. Another process, the "biological pump," involves the uptake of dissolved CO_2 by marine vegetation and phytoplankton (small free-floating photosynthetic organisms) living in the upper ocean or by other marine organisms that use CO_2 to build skeletons and other structures made of calcium carbonate ($CaCO_3$). As these organisms expire and fall to the ocean floor, the carbon they contain is transported downward and eventually buried at depth. A long-term balance between these natural sources and sinks leads to the background, or natural, level of CO_2 in the atmosphere.

In contrast, human activities increase atmospheric CO_2 levels primarily through the burning of fossil fuels—principally oil and coal and secondarily natural gas, for use in transportation, heating, and the generation of electrical power—and through the production of cement. Other anthropogenic sources include the burning of forests and the clearing of land. Anthropogenic emissions currently account for the annual release of about 7 gigatons (7 billion tons) of carbon into the atmosphere. Anthropogenic emissions are equal to approximately 3 percent of the total emissions of CO_2 by natural sources, and this amplified carbon load from human activities far exceeds the offsetting capacity of natural sinks (by perhaps as much as 2–3 gigatons per year).

Deforestation Smoldering remains of a plot of deforested land in the Amazon Rainforest. Annually, it is estimated that net global deforestation accounts for about two gigatons of carbon emissions to the atmosphere.

CO_2 consequently accumulated in the atmosphere at an average rate of 1.4 ppm per year between 1959 and 2006 and roughly 2.0 ppm per year between 2006 and 2018. Overall, this rate of accumulation has been linear (that is, uniform over time). However, certain current sinks, such as the oceans, could become sources in the future. This may lead to a situation in which the concentration of atmospheric CO_2 builds at an exponential rate (that is, its rate of increase is also increasing).

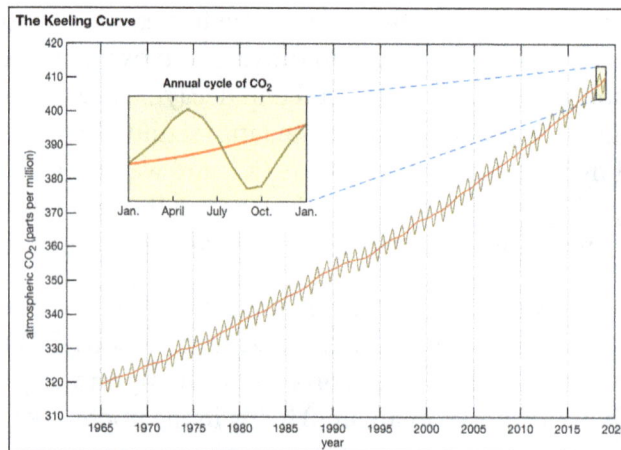

The Keeling Curve, tracks changes in the concentration of carbon dioxide (CO_2) in Earth's atmosphere at a research station on Mauna Loa. Although these concentrations experience small seasonal fluctuations, the overall trend shows that CO_2 is increasing in the atmosphere.

The natural background level of carbon dioxide varies on timescales of millions of years because of slow changes in outgassing through volcanic activity. For example, roughly 100 million years ago, during the Cretaceous Period (145 million to 66 million years ago), CO_2 concentrations appear to have been several times higher than they are today (perhaps close to 2,000 ppm). Over the past 700,000 years, CO_2 concentrations have varied over a far smaller range (between roughly 180 and 300 ppm) in association with the same Earth orbital effects linked to the coming and going of the Pleistocene ice ages. By the early 21st century, CO_2 levels had reached 384 ppm, which is approximately 37 percent above the natural background level of roughly 280 ppm that existed at the beginning of the Industrial Revolution. Atmospheric CO_2 levels continued to increase, and by 2018 they had reached 410 ppm. Such levels are believed to be the highest in at least 800,000 years according to ice core measurements and may be the highest in at least 5 million years according to other lines of evidence.

Radiative forcing caused by carbon dioxide varies in an approximately logarithmic fashion with the concentration of that gas in the atmosphere. The logarithmic relationship occurs as the result of a saturation effect wherein it becomes increasingly difficult, as CO_2 concentrations increase, for additional CO_2 molecules to further influence the "infrared window" (a certain narrow band of wavelengths in the infrared region that is not absorbed by atmospheric gases). The logarithmic relationship predicts that the surface warming potential will rise by roughly the same amount for each doubling of CO_2 concentration. At current rates of fossil fuel use, a doubling of CO_2 concentrations over preindustrial levels is expected to take place by the middle of the 21st century (when CO_2 concentrations are projected to reach 560 ppm). A doubling of CO_2 concentrations would represent an increase of roughly 4 watts per square metre of radiative forcing. Given typical estimates of "climate sensitivity" in the absence of any offsetting factors, this energy increase would lead to a warming of 2 to 5 °C (3.6 to 9 °F) over preindustrial times. The total radiative forcing by anthropogenic CO_2 emissions since the beginning of the industrial age is approximately 1.66 watts per square metre.

Methane

Methane (CH_4) is the second most important greenhouse gas. CH_4 is more potent than CO_2 because the radiative forcing produced per molecule is greater. In addition, the infrared window is less saturated in the range of wavelengths of radiation absorbed by CH_4, so more molecules may fill in the region. However, CH_4 exists in far lower concentrations than CO_2 in the atmosphere, and its concentrations by volume in the atmosphere are generally measured in parts per billion (ppb) rather than ppm. CH_4 also has a considerably shorter residence time in the atmosphere than CO_2 (the residence time for CH_4 is roughly 10 years, compared with hundreds of years for CO_2).

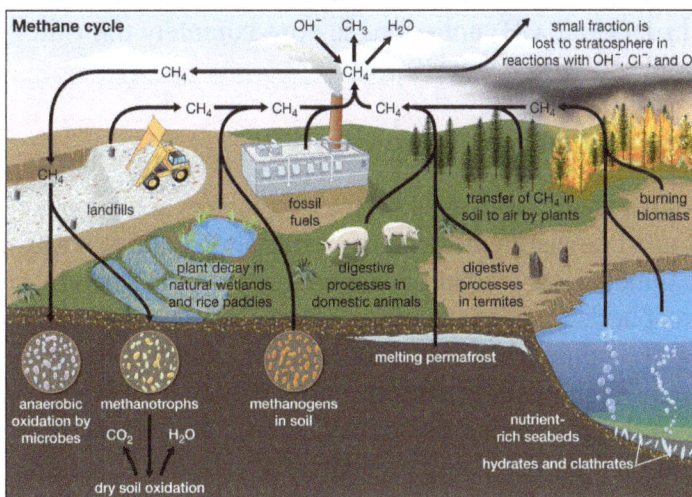

Methane cycle.

Natural sources of methane include tropical and northern wetlands, methane-oxidizing bacteria that feed on organic material consumed by termites, volcanoes, seepage vents of the seafloor in regions rich with organic sediment, and methane hydrates trapped along the continental shelves of the oceans and in polar permafrost. The primary natural sink for methane is the atmosphere itself, as methane reacts readily with the hydroxyl radical (·OH) within the troposphere to form CO_2 and water vapour (H_2O). When CH_4 reaches the stratosphere, it is destroyed. Another natural sink is soil, where methane is oxidized by bacteria.

As with CO_2, human activity is increasing the CH_4 concentration faster than it can be offset by natural sinks. Anthropogenic sources currently account for approximately 70 percent of total annual emissions, leading to substantial increases in concentration over time. The major anthropogenic sources of atmospheric CH_4 are rice cultivation, livestock farming, the burning of coal and natural gas, the combustion of biomass, and the decomposition of organic matter in landfills. Future trends are particularly difficult to anticipate. This is in part due to an incomplete understanding of the climate feedbacks associated with CH_4 emissions. In addition it is difficult to predict how, as human populations grow, possible changes in livestock raising, rice cultivation, and energy utilization will influence CH_4 emissions.

It is believed that a sudden increase in the concentration of methane in the atmosphere was responsible for a warming event that raised average global temperatures by 4–8 °C (7.2–14.4 °F) over a few thousand years during the so-called Paleocene-Eocene Thermal Maximum, or PETM. This episode took place roughly 55 million years ago, and the rise in CH_4 appears to have been related to a massive volcanic eruption that interacted with methane-containing flood deposits. As a result, large amounts of gaseous CH_4 were injected into the atmosphere. It is difficult to know precisely how high these concentrations were or how long they persisted. At very high concentrations, residence times of CH_4 in the atmosphere can become much greater than the nominal 10-year residence time that applies today. Nevertheless, it is likely that these concentrations reached several ppm during the PETM.

Methane concentrations have also varied over a smaller range (between roughly 350 and 800 ppb) in association with the Pleistocene ice age cycles. Preindustrial levels of CH_4 in the atmosphere were approximately 700 ppb, whereas levels exceeded 1,867 ppb in late 2018. (These concentrations are well above the natural levels observed for at least the past 650,000 years.) The net radiative forcing by anthropogenic CH_4 emissions is approximately 0.5 watt per square metre—or roughly one-third the radiative forcing of CO_2.

Surface-level Ozone and other Compounds

The next most significant greenhouse gas is surface, or low-level, ozone (O_3). Surface O_3 is a result of air pollution; it must be distinguished from naturally occurring stratospheric O_3, which has a very different role in the planetary radiation balance. The primary natural source of surface O_3 is the subsidence of stratospheric O_3 from the upper atmosphere. In contrast, the primary anthropogenic source of surface O_3 is photochemical reactions involving the atmospheric pollutant carbon monoxide (CO). The best estimates of the natural concentration of surface O_3 are 10 ppb, and the net radiative forcing due to anthropogenic emissions of surface O_3 is approximately 0.35 watt per square metre. Ozone concentrations can rise above unhealthy levels (that is, conditions where concentrations meet or exceed 70 ppb for eight hours or longer) in cities prone to photochemical smog.

Nitrous Oxides and Fluorinated Gases

Additional trace gases produced by industrial activity that have greenhouse properties include nitrous oxide (N_2O) and fluorinated gases (halocarbons), the latter including sulfur hexafluoride, hydrofluorocarbons (HFCs), and perfluorocarbons (PFCs). Nitrous oxide is responsible for 0.16 watt per square metre radiative forcing, while fluorinated gases are collectively responsible for 0.34 watt per square metre. Nitrous oxides have small background concentrations due to natural biological reactions in soil and water, whereas the fluorinated gases owe their existence almost entirely to industrial sources.

Aerosols

The production of aerosols represents an important anthropogenic radiative forcing of climate. Collectively, aerosols block—that is, reflect and absorb—a portion of incoming solar radiation, and this creates a negative radiative forcing. Aerosols are second only to greenhouse gases in relative importance in their impact on near-surface air temperatures. Unlike the decade-long residence times of the "well-mixed" greenhouse gases, such as CO_2 and CH_4, aerosols are readily flushed out of the atmosphere within days, either by rain or snow (wet deposition) or by settling out of the air (dry deposition). They must therefore be continually generated in order to produce a steady effect on radiative forcing. Aerosols have the ability to influence climate directly by absorbing or reflecting incoming solar radiation, but they can also produce indirect effects on climate by modifying cloud formation or cloud properties. Most aerosols serve as condensation nuclei (surfaces upon which water vapour can condense to form clouds); however, darker-coloured aerosols may hinder cloud formation by absorbing sunlight and heating up the surrounding air. Aerosols can be transported thousands of kilometres from their sources of origin by winds and upper-level circulation in the atmosphere.

Perhaps the most important type of anthropogenic aerosol in radiative forcing is sulfate aerosol. It is produced from sulfur dioxide (SO_2) emissions associated with the burning of coal and oil. Since the late 1980s, global emissions of SO_2 have decreased from about 151.5 million tonnes (167.0 million tons) to less than 100 million tonnes (110.2 million tons) of sulfur per year.

Nitrate aerosol is not as important as sulfate aerosol, but it has the potential to become a significant source of negative forcing. One major source of nitrate aerosol is smog (the combination of ozone with oxides of nitrogen in the lower atmosphere) released from the incomplete burning of fuel in internal-combustion engines. Another source is ammonia (NH_3), which is often used in fertilizers or released by the burning of plants and other organic materials. If greater amounts of atmospheric nitrogen are converted to ammonia and agricultural ammonia emissions continue to increase as projected, the influence of nitrate aerosols on radiative forcing is expected to grow.

Both sulfate and nitrate aerosols act primarily by reflecting incoming solar radiation, thereby reducing the amount of sunlight reaching the surface. Most aerosols, unlike greenhouse gases, impart a cooling rather than warming influence on Earth's surface. One prominent exception is carbonaceous aerosols such as carbon black or soot, which are produced by the burning of fossil fuels and biomass. Carbon black tends to absorb rather than reflect incident solar radiation, and so it has a warming impact on the lower atmosphere, where it resides. Because of its absorptive properties, carbon black is also capable of having an additional indirect effect on climate. Through its deposition in snowfall, it can decrease the albedo of snow cover. This reduction in the amount of solar radiation reflected back to space by snow surfaces creates a minor positive radiative forcing.

Natural forms of aerosol include windblown mineral dust generated in arid and semiarid regions and sea salt produced by the action of waves breaking in the ocean. Changes to wind patterns as a result of climate modification could alter the emissions of these aerosols. The influence of climate change on regional patterns of aridity could shift both the sources and the destinations of dust clouds. In addition, since the concentration of sea salt aerosol, or sea aerosol, increases with the strength of the winds near the ocean surface, changes in wind speed due to global warming and climate change could influence the concentration of sea salt aerosol. For example, some studies suggest that climate change might lead to stronger winds over parts of the North Atlantic Ocean. Areas with stronger winds may experience an increase in the concentration of sea salt aerosol.

Other natural sources of aerosols include volcanic eruptions, which produce sulfate aerosol, and biogenic sources (e.g., phytoplankton), which produce dimethyl sulfide (DMS). Other important biogenic aerosols, such as terpenes, are produced naturally by certain kinds of trees or other plants. For example, the dense forests of the Blue Ridge Mountains of Virginia in the United States emit terpenes during the summer months, which in turn interact with the high humidity and warm temperatures to produce a natural photochemical smog. Anthropogenic pollutants such as nitrate and ozone, both of which serve as precursor molecules for the generation of biogenic aerosol, appear to have increased the rate of production of these aerosols severalfold. This process appears to be responsible for some of the increased aerosol pollution in regions undergoing rapid urbanization.

Human activity has greatly increased the amount of aerosol in the atmosphere compared with the background levels of preindustrial times. In contrast to the global effects of greenhouse gases, the impact of anthropogenic aerosols is confined primarily to the Northern Hemisphere, where most of the world's industrial activity occurs. The pattern of increases in anthropogenic aerosol over time is also somewhat different from that of greenhouse gases. During the middle of the 20th century, there was a substantial increase in aerosol emissions. This appears to have been at least partially responsible for a cessation of surface warming that took place in the Northern Hemisphere from the 1940s through the 1970s. Since that time, aerosol emissions have leveled off due to antipollution measures undertaken in the industrialized countries since the 1960s. Aerosol emissions may rise in the future, however, as a result of the rapid emergence of coal-fired electric power generation in China and India.

The total radiative forcing of all anthropogenic aerosols is approximately −1.2 watts per square metre. Of this total, −0.5 watt per square metre comes from direct effects (such as the reflection of solar energy back into space), and −0.7 watt per square metre comes from indirect effects (such as the influence of aerosols on cloud formation). This negative radiative forcing represents an offset of roughly 40 percent from the positive radiative forcing caused by human activity. However, the relative uncertainty in aerosol radiative forcing (approximately 90 percent) is much greater than that of greenhouse gases. In addition, future emissions of aerosols from human activities, and the influence of these emissions on future climate change, are not known with any certainty. Nevertheless, it can be said that, if concentrations of anthropogenic aerosols continue to decrease as they have since the 1970s, a significant offset to the effects of greenhouse gases will be reduced, opening future climate to further warming.

Land-use Change

There are a number of ways in which changes in land use can influence climate. The most direct influence is through the alteration of Earth's albedo, or surface reflectance. For example, the replacement of forest by cropland and pasture in the middle latitudes over the past several centuries has led to an increase in albedo, which in turn has led to greater reflection of incoming solar radiation in those regions. This replacement of forest by agriculture has been associated with a change in global average radiative forcing of approximately −0.2 watt per square metre since 1750. In Europe and other major agricultural regions, such land-use conversion began more than 1,000 years ago and has proceeded nearly to completion. For Europe, the negative radiative forcing due to land-use change has probably been substantial, perhaps approaching −5 watts per square metre. The influence of early land use on radiative forcing may help to explain a long period of cooling in Europe that followed a period of relatively mild conditions roughly 1,000 years ago. It is generally believed that the mild temperatures of this "medieval warm period," which was followed by a long period of cooling, rivaled those of 20th-century Europe.

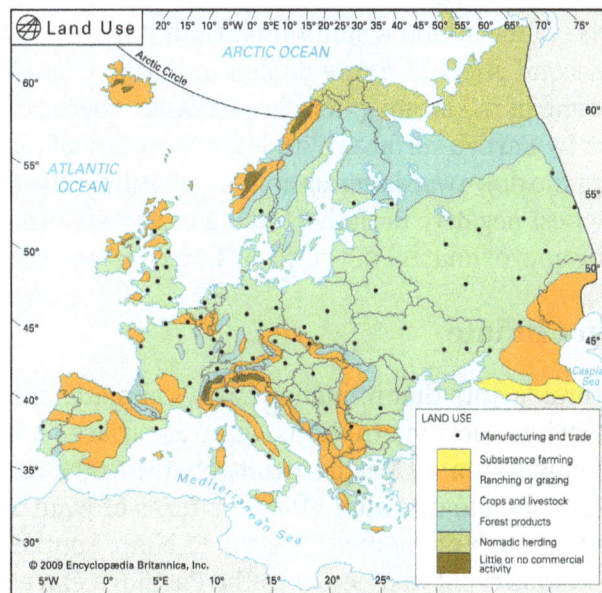

Land use in Europe.

Land-use changes can also influence climate through their influence on the exchange of heat between Earth's surface and the atmosphere. For example, vegetation helps to facilitate the evaporation of water into the atmosphere through evapotranspiration. In this process, plants take up liquid water from the soil through their root systems. Eventually this water is released through transpiration into the atmosphere, as water vapour through the stomata in leaves. While deforestation generally leads to surface cooling due to the albedo factor the land surface may also be warmed as a result of the release of latent heat by the evapotranspiration process. The relative importance of these two factors, one exerting a cooling effect and the other a warming effect, varies by both season and region. While the albedo effect is likely to dominate in middle latitudes, especially during the period from autumn through spring, the evapotranspiration effect may dominate during the summer in the midlatitudes and year-round in the tropics. The latter case is particularly important in assessing the potential impacts of continued tropical deforestation.

The rate at which tropical regions are deforested is also relevant to the process of carbon sequestration, the long-term storage of carbon in underground cavities and biomass rather than in the atmosphere. By removing carbon from the atmosphere, carbon sequestration acts to mitigate global warming. Deforestation contributes to global warming, as fewer plants are available to take up carbon dioxide from the atmosphere. In addition, as fallen trees, shrubs, and other plants are burned or allowed to slowly decompose, they release as carbon dioxide the carbon they stored during their lifetimes. Furthermore, any land-use change that influences the amount, distribution, or type of vegetation in a region can affect the concentrations of biogenic aerosols, though the impact of such changes on climate is indirect and relatively minor.

Stratospheric Ozone Depletion

Since the 1970s the loss of ozone (O_3) from the stratosphere has led to a small amount of negative radiative forcing of the surface. This negative forcing represents a competition between two distinct effects caused by the fact that ozone absorbs solar radiation. In the first case, as ozone levels in the stratosphere are depleted, more solar radiation reaches Earth's surface. In the absence of any other influence, this rise in insolation would represent a positive radiative forcing of the surface. However, there is a second effect of ozone depletion that is related to its greenhouse properties. As the amount of ozone in the stratosphere is decreased, there is also less ozone to absorb longwave radiation emitted by Earth's surface. With less absorption of radiation by ozone, there is a corresponding decrease in the downward reemission of radiation. This second effect overwhelms the first and results in a modest negative radiative forcing of Earth's surface and a modest cooling of the lower stratosphere by approximately 0.5 °C (0.9 °F) per decade since the 1970s.

Natural Influences on Climate

There are a number of natural factors that influence Earth's climate. These factors include external influences such as explosive volcanic eruptions, natural variations in the output of the Sun, and slow changes in the configuration of Earth's orbit relative to the Sun. In addition, there are natural oscillations in Earth's climate that alter global patterns of wind circulation, precipitation, and surface temperatures. One such phenomenon is the El Niño/Southern Oscillation (ENSO), a coupled atmospheric and oceanic event that occurs in the Pacific Ocean every three to seven years. In addition, the Atlantic Multidecadal Oscillation (AMO) is a similar phenomenon that occurs over decades in the North Atlantic Ocean. Other types of oscillatory behaviour that produce dramatic shifts in climate may occur across timescales of centuries and millennia.

Volcanic Aerosols

Explosive volcanic eruptions have the potential to inject substantial amounts of sulfate aerosols into the lower stratosphere. In contrast to aerosol emissions in the lower troposphere, aerosols that enter the stratosphere may remain for several years before settling out, because of the relative absence of turbulent motions there. Consequently, aerosols from explosive volcanic eruptions have the potential to affect Earth's climate. Less-explosive eruptions, or eruptions that are less vertical in orientation, have a lower potential for substantial climate impact. Furthermore, because of large-scale circulation patterns within the stratosphere, aerosols injected within tropical regions tend to spread out over the globe, whereas aerosols injected within midlatitude and polar regions tend to remain confined to the middle and high latitudes of that hemisphere. Tropical eruptions, therefore, tend to have a greater climatic impact than eruptions occurring toward the poles. In 1991 the moderate eruption of Mount Pinatubo in the Philippines provided a peak forcing of approximately −4 watts per square metre and cooled the climate by about 0.5 °C (0.9 °F) over the following few years. By comparison, the 1815 Mount Tambora eruption in present-day Indonesia, typically implicated for the 1816 "year without a summer" in Europe and North America, is believed to have been associated with a radiative forcing of approximately −6 watts per square metre.

While in the stratosphere, volcanic sulfate aerosol actually absorbs longwave radiation emitted by Earth's surface, and absorption in the stratosphere tends to result in a cooling of the troposphere below. This vertical pattern of temperature change in the atmosphere influences the behaviour of winds in the lower atmosphere, primarily in winter. Thus, while there is essentially a global cooling effect for the first few years following an explosive volcanic eruption, changes in the winter patterns of surface winds may actually lead to warmer winters in some areas, such as Europe. There is also evidence that volcanic eruptions may influence other climate phenomena such as ENSO.

Variations in Solar Output

Direct measurements of solar irradiance, or solar output, have been available from satellites only since the late 1970s. These measurements show a very small peak-to-peak variation in solar irradiance (roughly 0.1 percent of the 1,366 watts per square metre received at the top of the atmosphere, for approximately 1.4 watts per square metre). However, indirect measures of solar activity are available from historical sunspot measurements dating back through the early 17th century. Attempts have been made to reconstruct graphs of solar irradiance variations from historical sunspot data by calibrating them against the measurements from modern satellites. However, since the modern measurements span only a few of the most recent 11-year solar cycles, estimates of solar output variability on 100-year and longer timescales are poorly correlated. Different assumptions regarding the relationship between the amplitudes of 11-year solar cycles and long-period solar output changes can lead to considerable differences in the resulting solar reconstructions. These differences in turn lead to fairly large uncertainty in estimating positive forcing by changes in solar irradiance since 1750. (Estimates range from 0.06 to 0.3 watt per square metre.) Even more challenging, given the lack of any modern analog, is the estimation of solar irradiance during the so-called Maunder Minimum, a period lasting from the mid-17th century to the early 18th century when very few sunspots were observed. While it is likely that solar irradiance was reduced at this time, it is difficult to calculate by how much. However, additional proxies of solar output exist that match reasonably well with the sunspot-derived records following the Maunder Minimum; these may be used as crude estimates of the solar irradiance variations.

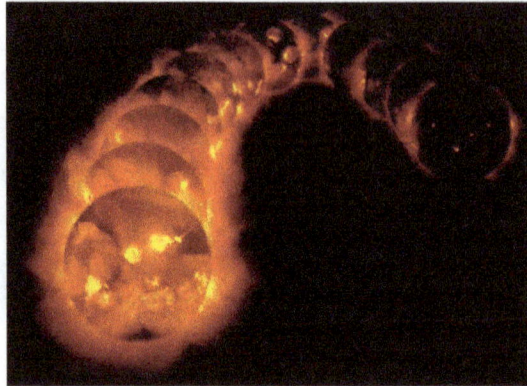

Twelve solar X-ray images. The solar coronal brightness decreases by a factor of about 100 during a solar cycle as the Sun goes from an "active" state (left) to a less active state (right).

In theory it is possible to estimate solar irradiance even farther back in time, over at least the past millennium, by measuring levels of cosmogenic isotopes such as carbon-14 and beryllium-10. Cosmogenic isotopes are isotopes that are formed by interactions of cosmic rays with atomic nuclei in the atmosphere and that subsequently fall to Earth, where they can be measured in the annual layers found in ice cores. Since their production rate in the upper atmosphere is modulated by changes in solar activity, cosmogenic isotopes may be used as indirect indicators of solar irradiance. However, as with the sunspot data, there is still considerable uncertainty in the amplitude of past solar variability implied by these data.

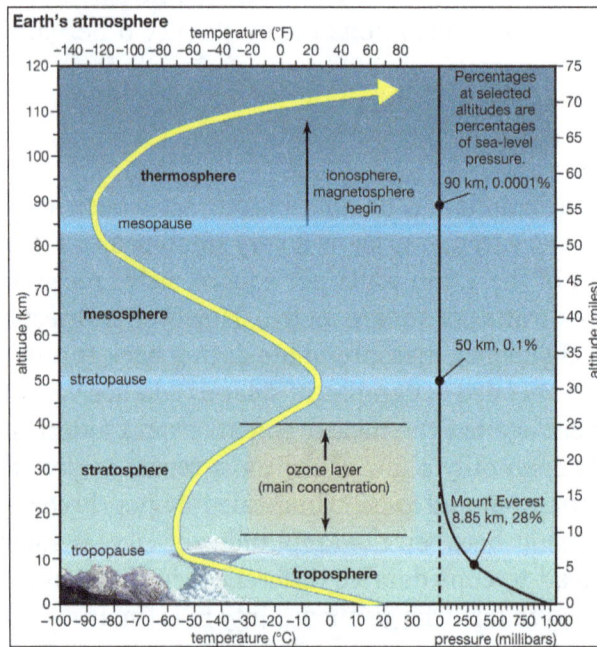

The layers of Earth's atmosphere. The yellow line shows the response of air temperature to increasing height.

Solar forcing also affects the photochemical reactions that manufacture ozone in the stratosphere. Through this modulation of stratospheric ozone concentrations, changes in solar irradiance (particularly in the ultraviolet portion of the electromagnetic spectrum) can modify how both shortwave and longwave radiation in the lower stratosphere are absorbed. As a result, the vertical temperature profile of the atmosphere can change, and this change can in turn influence phenomena such as the strength of the winter jet streams.

Variations in Earth's Orbit

On timescales of tens of millennia, the dominant radiative forcing of Earth's climate is associated with slow variations in the geometry of Earth's orbit about the Sun. These variations include the precession of the equinoxes (that is, changes in the timing of summer and winter), occurring on a roughly 26,000-year timescale; changes in the tilt angle of Earth's rotational axis relative to the plane of Earth's orbit around the Sun, occurring on a roughly 41,000-year timescale; and changes in the eccentricity (the departure from a perfect circle) of Earth's orbit around the Sun, occurring on a roughly 100,000-year timescale. Changes in eccentricity slightly influence the mean annual solar radiation at the top of Earth's atmosphere, but the primary influence of all the orbital variations listed above is on the seasonal and latitudinal distribution of incoming solar radiation over Earth's surface. The major ice ages of the Pleistocene Epoch were closely related to the influence of these variations on summer insolation at high northern latitudes. Orbital variations thus exerted a primary control on the extent of continental ice sheets. However, Earth's orbital changes are generally believed to have had little impact on climate over the past few millennia, and so they are not considered to be significant factors in present-day climate variability.

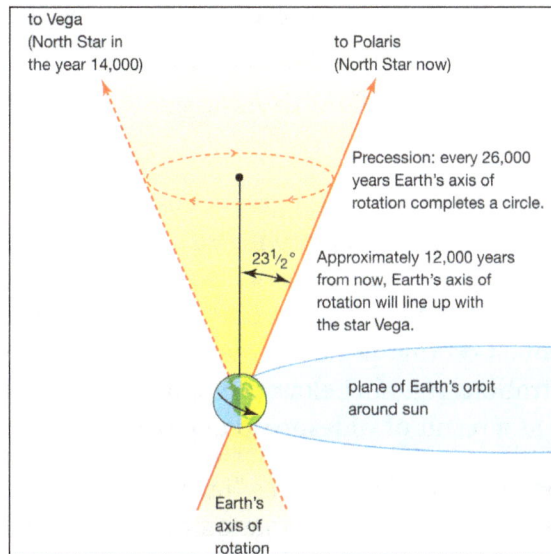

In the figure, Earth's axis of rotation itself rotates, or precesses, completing one circle every 26,000 years. Consequently, Earth's North Pole points toward different stars (and sometimes toward empty space) as it travels in this circle. This precession is so slow that it is not noticeable in a person's lifetime, though astronomers must consider its effect when studying ancient sites such as Stonehenge.

Potential Effects of Global Warming

The path of future climate change will depend on what courses of action are taken by society—in particular the emission of greenhouse gases from the burning of fossil fuels. A range of alternative emissions scenarios known as representative concentration pathways (RCPs) were proposed by the IPCC in the Fifth Assessment Report (AR5), which was published in 2014, to examine potential future climate changes. The scenarios depend on various assumptions concerning future rates of human population growth, economic development, energy demand, technological advancement, and other factors.

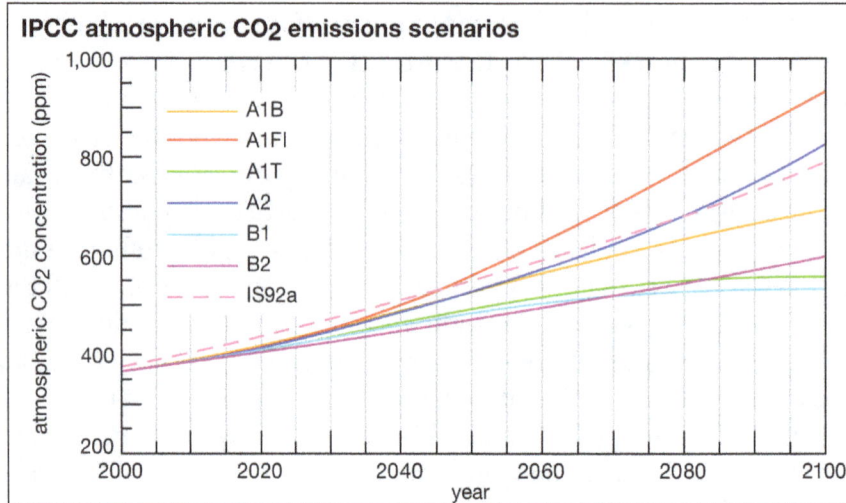

Carbon Dioxide: Global Warming Scenarios Graph of the predicted increase in the concentration of carbon dioxide (CO_2) in Earth's atmosphere according to a series of climate change scenarios that assume different levels of economic development, population growth, and fossil fuel use.

The results of each scenario in the IPCC's Fourth Assessment Report (2007) are depicted in the graph.

The AR5 scenario with the smallest increases in greenhouse gases is RCP 2.6, which denotes the net radiative forcing by 2100 in watts per square metre (a doubling of CO_2 concentrations from preindustrial values of 280 ppm to 560 ppm represents roughly 3.7 watts per square metre). RCP 2.6 assumes substantial improvements in energy efficiency, a rapid transition away from fossil fuel energy, and a global population that peaks at roughly nine billion people in the 21st century. In that scenario CO_2 concentrations remain below 450 ppm and actually fall toward the end of the century (to about 420 ppm) as a result of widespread deployment of carbon-capture technology.

Scenario RCP 8.5, by contrast, might be described as "business as usual." It reflects the assumption of an energy-intensive global economy, high population growth, and a reduced rate of technological development. CO_2 concentrations are more than three times greater than preindustrial levels (roughly 936 ppm) by 2100 and continue to grow thereafter. RCP 4.5 and RCP 6.0 envision intermediate policy choices, resulting in stabilization by 2100 of CO_2 concentrations at 538 and 670 ppm, respectively. In all those scenarios, the cooling effect of industrial pollutants such as sulfate particulates, which have masked some of the past century's warming, is assumed to decline to near zero by 2100 because of policies restricting their industrial production.

Simulations of Future Climate Change

The differences between the various simulations arise from disparities between the various climate models used and from assumptions made by each emission scenario. For example, best estimates of the predicted increases in global surface temperature between the years 2000 and 2100 range from about 0.3 to 4.8 °C (0.5 to 8.6 °F), depending on which emission scenario is assumed and which climate model is used. Relative to preindustrial (i.e., 1750–1800) temperatures, these estimates reflect an overall warming of the globe of 1.4 to 5.0 °C (2.5 to 9.0 °F). These projections are conservative in that they do not take into account potential positive carbon cycle feedbacks. Only the lower-end emissions scenario RCP 2.6 has a reasonable chance (roughly 50 percent) of holding additional global surface warming by 2100 to less than 2.0 °C (3.6 °F)—a level considered by many scientists to be the threshold above which pervasive and extreme climatic effects will occur.

Patterns of Warming

The greatest increase in near-surface air temperature is projected to occur over the polar region of the Northern Hemisphere because of the melting of sea ice and the associated reduction in surface albedo. Greater warming is predicted over land areas than over the ocean. Largely due to the delayed warming of the oceans and their greater specific heat, the Northern Hemisphere—with less than 40 percent of its surface area covered by water—is expected to warm faster than the Southern Hemisphere. Some of the regional variation in predicted warming is expected to arise from changes to wind patterns and ocean currents in response to surface warming. For example, the warming of the region of the North Atlantic Ocean just south of Greenland is expected to be slight. This anomaly is projected to arise from a weakening of warm northward ocean currents combined with a shift in the jet stream that will bring colder polar air masses to the region.

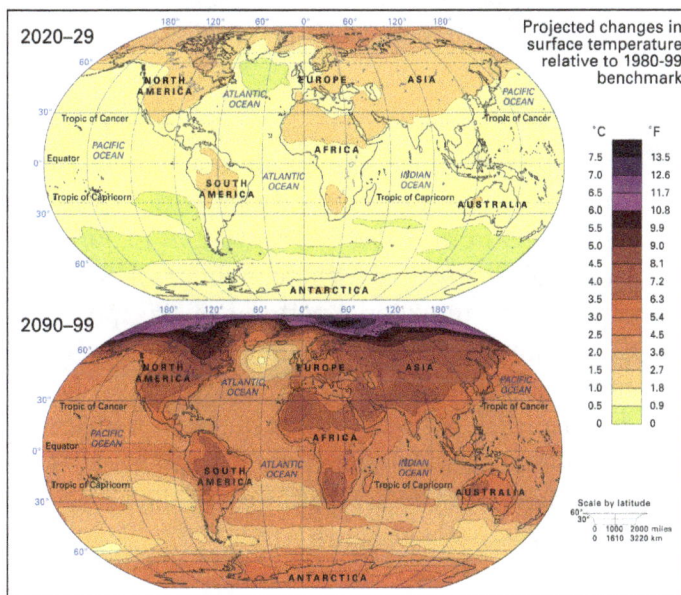

Projected changes in mean surface temperatures by the late 21st century according to the A1B climate change scenario. All values for the period 2090–99 are shown relative to the mean temperature values for the period 1980–99.

Precipitation Patterns

The climate changes associated with global warming are also projected to lead to changes in precipitation patterns across the globe. Increased precipitation is predicted in the polar and subpolar regions, whereas decreased precipitation is projected for the middle latitudes of both hemispheres as a result of the expected poleward shift in the jet streams. Whereas precipitation near the Equator is predicted to increase, it is thought that rainfall in the subtropics will decrease. Both phenomena are associated with a forecasted strengthening of the tropical Hadley cell pattern of atmospheric circulation.

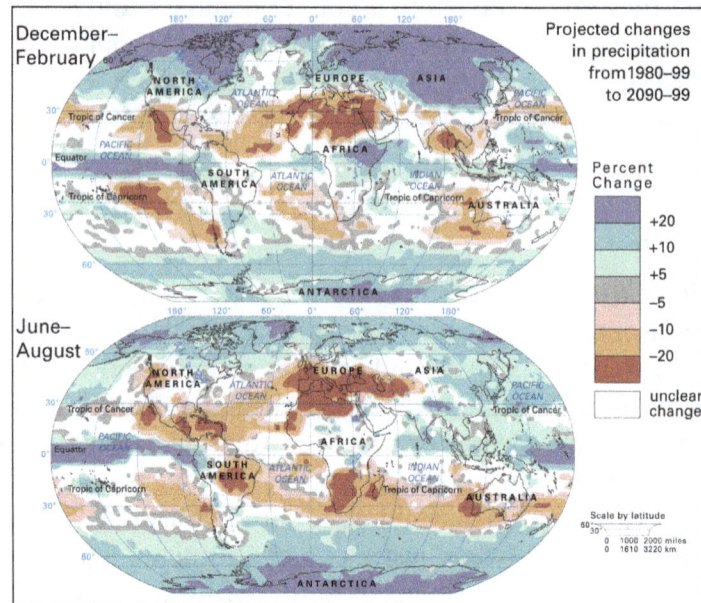

Projected changes in mean precipitation by the late 21st century according to the A1B climate change scenario. All values for the period 2090–99 are shown as a percentage relative to the mean precipitation values for the period 1980–99.

Regional Predictions

Regional predictions of future climate change remain limited by uncertainties in how the precise patterns of atmospheric winds and ocean currents will vary with increased surface warming. For example, some uncertainty remains in how the frequency and magnitude of El Niño/Southern Oscillation (ENSO) events will adjust to climate change. Since ENSO is one of the most prominent sources of interannual variations in regional patterns of precipitation and temperature, any uncertainty in how it will change implies a corresponding uncertainty in certain regional patterns of climate change. For example, increased El Niño activity would likely lead to more winter precipitation in some regions, such as the desert southwest of the United States. This might offset the drought predicted for those regions, but at the same time it might lead to less precipitation in other regions. Rising winter precipitation in the desert southwest of the United States might exacerbate drought conditions in locations as far away as South Africa.

Ice Melt and Sea Level Rise

A warming climate holds important implications for other aspects of the global environment. Because of the slow process of heat diffusion in water, the world's oceans are likely to continue to warm for several centuries in response to increases in greenhouse concentrations that have taken place so far. The combination of seawater's thermal expansion associated with this warming and the melting of mountain glaciers is predicted to lead to an increase in global sea level of 0.45–0.82 metre (1.4–2.7 feet) by 2100 under the RCP 8.5 emissions scenario. However, the actual rise in sea level could be considerably greater than this. It is probable that the continued warming of Greenland will cause its ice sheet to melt at accelerated rates. In addition, this level of surface warming may also melt the ice sheet of West Antarctica. Paleoclimatic evidence suggests that an additional 2 °C (3.6 °F) of warming could lead to the ultimate destruction of the Greenland Ice Sheet, an event that would add another 5 to 6 metres (16 to 20 feet) to predicted sea level rise. Such an increase would submerge a substantial number of islands and lowland regions. Coastal lowland regions vulnerable to sea level rise include substantial parts of the U.S. Gulf Coast and Eastern Seaboard (including roughly the lower third of Florida), much of the Netherlands and Belgium (two of the European Low Countries), and heavily populated tropical areas such as Bangladesh. In addition, many of the world's major cities—such as Tokyo, New York, Mumbai, Shanghai, and Dhaka—are located in lowland regions vulnerable to rising sea levels. With the loss of the West Antarctic ice sheet, additional sea level rise would approach 10.5 metres (34 feet).

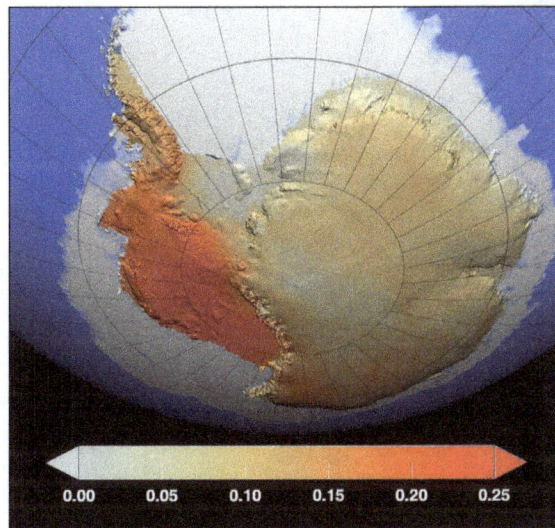

NASA image showing locations on Antarctica where temperatures had increased between 1959 and 2009. Red represents areas where temperatures had increased the most over the period, particularly in West Antarctica, while dark blue represents areas with a lesser degree of warming. Temperature changes are measured in degrees Celsius.

While the current generation of models predicts that such global sea level changes might take several centuries to occur, it is possible that the rate could accelerate as a result of processes that tend to hasten the collapse of ice sheets. One such process is the development of moulins—large vertical shafts in the ice that allow surface meltwater to penetrate to the base of the ice sheet. A second process involves the vast ice shelves off Antarctica that buttress the grounded continental ice sheet of Antarctica's interior. If those ice shelves collapse, the continental ice sheet could become unstable, slide rapidly toward the ocean, and melt, thereby further increasing mean sea level. Thus far, neither process has been incorporated into the theoretical models used to predict sea level rise.

Ocean Circulation Changes

Another possible consequence of global warming is a decrease in the global ocean circulation system known as the "thermohaline circulation" or "great ocean conveyor belt." This system involves the sinking of cold saline waters in the subpolar regions of the oceans, an action that helps to drive warmer surface waters poleward from the subtropics. As a result of this process, a warming influence is carried to Iceland and the coastal regions of Europe that moderates the climate in those regions. Some scientists believe that global warming could shut down this ocean current system by creating an influx of fresh water from melting ice sheets and glaciers into the subpolar North Atlantic Ocean. Since fresh water is less dense than saline water, a significant intrusion of fresh water would lower the density of the surface waters and thus inhibit the sinking motion that drives the large-scale thermohaline circulation. It has also been speculated that, as a consequence of large-scale surface warming, such changes could even trigger colder conditions in regions surrounding the North Atlantic. Experiments with modern climate models suggest that such an event would be unlikely. Instead, a moderate weakening of the thermohaline circulation might occur that would lead to a dampening of surface warming—rather than actual cooling—in the higher latitudes of the North Atlantic Ocean.

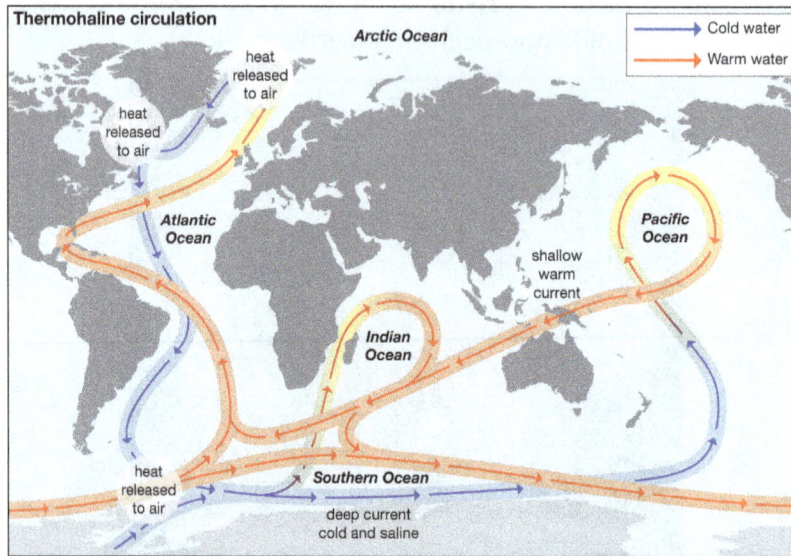

Thermohaline circulation transports and mixes the water of the oceans. In the process it transports heat, which influences regional climate patterns. The density of seawater is determined by the temperature and salinity of a volume of seawater at a particular location. The difference in density between one location and another drives the thermohaline circulation.

Tropical Cyclones

One of the more controversial topics in the science of climate change involves the impact of global warming on tropical cyclone activity. It appears likely that rising tropical ocean temperatures associated with global warming will lead to an increase in the intensity (and the associated destructive potential) of tropical cyclones. In the Atlantic a close relationship has been observed between rising ocean temperatures and a rise in the strength of hurricanes. Trends in the intensities of tropical cyclones in other regions, such as in the tropical Pacific and Indian oceans, are more uncertain due to a paucity of reliable long-term measurements.

While the warming of oceans favours increased tropical cyclone intensities, it is unclear to what extent rising temperatures affect the number of tropical cyclones that occur each year. Other factors, such as wind shear, could play a role. If climate change increases the amount of wind shear—a factor that discourages the formation of tropical cyclones—in regions where such storms tend to form, it might partially mitigate the impact of warmer temperatures. On the other hand, changes in atmospheric winds are themselves uncertain—because of, for example, uncertainties in how climate change will affect ENSO.

Global Dimming

Global dimming is defined as the decrease in the amounts of solar radiation reaching the surface of the Earth. The by-product of fossil fuels is tiny particles or pollutants which absorb solar energy and reflect back sunlight into space.

Various regions observe different levels of global dimming. Till now, the Southern Hemisphere has seen very small amounts of global dimming while Northern Hemisphere has witnessed more significant reductions, to the tune of 4-8%. Regions such as parts of Europe and North America has observed partial recovery from dimming while parts of China and India have experienced an increase in global dimming.

Causes of Global Dimming

Earlier it was thought that changes in the sun's luminosity cause global dimming but later it was realized that this was very small to explain the enormity of global dimming.

Aerosols have been found to be the major cause of global dimming. The burning of fossil fuels by industry and internal combustion engines emits by-products such as sulfur dioxide, soot, and ash. These together form particulate pollution—primarily called aerosols. Aerosols act as a precursor to global dimming in the following two ways:

- These particle matters enter the atmosphere and directly absorb solar energy and reflect radiation back into space before it reaches the planet's surface.

- Water droplets containing these air-borne particles form polluted clouds. These polluted clouds have a heavier and larger number of droplets. These changed properties of the cloud – such clouds are called 'brown clouds' – makes them more reflective.

Vapors emitted from the planes flying high in the sky called contrails are another cause of heat reflection and related global dimming.

Both global dimming and global warming have been happening all over the world and together they have caused severe changes in the rainfall patterns. It is also believed that it was global dimming behind the 1984 Saharan drought that killed millions of people in sub-Saharan Africa. Scientists believe that despite the cooling effect created by global dimming, the earth's temperature has increased by more than 1 deg. in the last century.

If global dimming wouldn't have happened, the temperature of this planet would be much higher and could have posed a serious effect o the lives of humans, plants, and animals.

Due to an increase in the burning of fossil fuels and pollution from various other sources, the heat is getting trapped into the earth's atmosphere. This heat has resulted in a marginal increase in earth's temperature. This is due to the fact that warming from greenhouse gases is getting offset by cooling effect from global dimming.

Effects of Global Dimming

Global dimming has devastating effects on the earth's environment and living beings. The pollutants causing global dimming also leads to acid rain, smog and respiratory diseases in humans.

Due to the reflection of solar energy and global dimming, the water in the northern hemisphere has become colder. This leads to slow evaporation and generation of lesser water droplets. This further causes a reduction in the amount of rain reaching certain parts of the globe, resulting in drought and famine situations. This has tragic consequences like miserable lives and deaths due to starvation.

It has now been determined that the drought and famine of The Sahel which has killed thousands of innocent people in sub-Saharan Africa during 1970s, was largely due to global dimming. A growing concern is the impact of global dimming on Asian monsoons which cause 50% of the world's annual rainfall. If this happens, then half of the world's population will be starving.

Global dimming is also believed to cause heat waves and runaway fires. Also, a decrease in sunlight or solar radiation will negatively impact the process of photosynthesis in plants.

Global dimming is thought to be counteracting the actual effect of carbon emissions on global warming. So, if efforts are made to reduce particulate emission causing global dimming, it will enhance global warming and increase the global temperatures to more than double. This will make planet Earth, almost uninhabitable. To prevent this situation, it is important that the emission of both greenhouse gases and particulate matter should be reduced simultaneously. This will balance out both the phenomena.

Global Dimming and Global Warming

Global warming and global dimming are opposite phenomena.

Global warming is defined as the increase in the atmospheric temperature. This largely caused by greenhouse gases (GHGs). Greenhouse gases produced from the burning of fossil fuels traps the infrared radiations. This heats up the earth's atmosphere.

Global dimming is the exact opposite. Fine particles such as aerosols, also produced as the by-product of fossil fuels burning, reflect away sunlight. This decreases the amount of solar radiation entering our planet. It produces a cooling effect.

Though both are opposite phenomena with contrasting effects but both are destructive for the planet. It is due to both global warming and global dimming that earth's temperature has increased less than what it should have been. Without global dimming, this planet would have turned to be too hot for all of us to survive. Both of them are dangerous and can prove fatal for our environment and need to be solved together. Solving each problem at a time could create conditions that may be harmful and may prove fatal for all of us.

References

* Climate-change: britannica.com, Retrieved 18 May, 2019

* Climate-impacts-ecosystems, climate-impacts: 19january2017snapshot.epa.gov, Retrieved 16 June, 2019

* Pollution-its-effects-on-climate: climatechange.insightconferences.com

* Link-between-climate-change-air-pollution, climate-change: activesustainability.com, Retrieved 17 May, 2019

* What-acid-rain, acidrain: epa.gov, Retrieved 16 April, 2019

* Causes-and-effects-of-global-dimming: conserve-energy-future.com, Retrieved 08 August, 2019

Chapter 5

Impacts of Human Activities on Climate

In order to completely understand climatology it is necessary to understand the impacts of human activities on climate. The following chapter elucidates the varied impacts associated with this area of study. Human activities such as deforestation, biomass burning, carbon dioxide emission, etc., have a negative impact on the climate.

Deforestation

Deforestation is an important factor in global climate change. It is well known that deforestation is a big problem in the world today, with hundreds and even thousands of vulnerable forest being cut down both for tinder and to make way for arable farmland for cows and other livestock. Not only this could lead to an increase in the rate of deforestation quickly and easily lead to the loss of many different species of plant, tree and animal – a lot of which are yet to be discovered – it could also have a devastating effect on the climate.

One of the main reasons for this is because forests all over the globe are, naturally, so-called 'carbon sinks', or areas of natural environment such as oceans that can take carbon dioxide from the atmosphere and convert it into oxygen that we and other animals can safely breathe. By cutting down huge areas of forest, therefore, without replacing the trees that we remove, we are causing an inadvertent change in the amount of carbon dioxide in the atmosphere, which can have a huge impact on the rest of the world.

Forests are vital for human and animal lives as they are home to millions of species, prevent soil erosion, play a crucial role in water cycle by returning water vapor back into the atmosphere, absorb greenhouse gases that fuel global warming, keep soil moist by blocking the sun, produce oxygen and absorb carbon dioxide.

Forests cover about 30% of the world's land area and large patches of forests are lost every year due to deforestation. If current rate of deforestation continue, there will be no more rainforests in 100 years. It is estimated that due to cutting and burning of forests every year, more than 1.5 billion tons of carbon dioxide are released into the atmosphere

Main Causes of Deforestation

There are many reasons that we cut down forests all over the world (many of which have no completely disappeared since thousands of years ago). These are listed below:

- Wood and Timber: This reason is obvious, and, of course, permissible. We need timber and wood to be able to build houses, furniture and other essential things that help us live the lives that we lead today. It is estimated that 500,000 hectares of forest all over the week are cut away each and every week by the logging industry, and that is not including the number of trees that are cut down by illegal loggers. Many of these trees never get replaced.

- Farmland: Strangely enough, the biggest threat to forests all over the world is not the need for timber or wood, it is the demand for arable farmland on which farmers can either grow crops or make way for cows to feed the population of the world. It has been suggested that as much as 80 percent of the farmland that was created between 1980 and 2000 meant that huge areas of forests had to be completely cut down. This means that more than half a million square miles of forest was cut down in the space of just twenty years, and more keeps getting cut down as our demand for meat and other foodstuffs creeps higher and higher.

- Urbanization: With more people being born every year and the older generations living for longer, there are more people on our planet than ever before. This means that we are starting to need more houses to accommodate them all, and houses need space. Some of the forest areas that are cut down are destroyed in order to make way for new houses and communities.

- Palm Oil: Palm oil that is not responsibly sourced has often come from a vast expanse of forest that has been cut down and not replaced with other trees or plants. Palm oil, as well as other consumer items, are harvested from huge areas of forest all over the globe and, as the demand for these sorts of products grows, so does the number of trees that are cut down.

- Paper: Until we start making a huge effort to start recycling paper, we are going to continue cutting down forests all around the world at an alarming rate. Paper is still common, even in today's world packed with technology and paperless machines. We still demand books, toilet paper, kitchen towels and other products that require us to cut down trees.

- Mining: Oil and mining companies require large amount of land to build plants. These plants come at a cost of chopping of thousands of plants and trees. Apart from that, roads and highways have to be built to connect cities and that requires trees to be cut down that come in the way.

Effects of Deforestation on Climate

With so many hundreds of thousands of square miles of forest being cut down each and every year (approximately 46 – 58 thousand), many of the biggest carbon sinks in the world are being shrunk drastically. This means that less of the carbon dioxide in our atmosphere is getting converted back into oxygen by photosynthesis, which is already having a huge impact on our climate. Below are few of the effects that deforestation can have on the climate.

- Global Warming: Carbon dioxide is a greenhouse gas, which means that it traps infra-red rays from the sun and keeps heat in the Earth's atmosphere. Whilst for the most part this is a good thing – without greenhouse gases it would be a lot more difficult for life to be sustained on the Earth's surface – more and more carbon dioxide in the atmosphere means more and more heat is being trapped, raising the average temperature of the world (otherwise known as global warming) which is having a huge number of knock-on effects.

- Weather Patterns: Water vapor is another greenhouse gas that keeps heat from the sun trapped in the Earth's atmosphere and helps maintain a temperature at which life can flourish on our planet. However, deforestation also affects the amount of this in the air. Studies have shown that deforestation has contributed to the amount of water vapor in the air and, over just a few years, the amount of water vapor present has increased by four percent. Besides increasing the temperature of the Earth's surface and atmosphere, this also has knock-on effects for weather patterns.

- Water Cycle: The water cycle is very important for not only us, but for species of animal and plant all over the globe. Trees contribute a great deal to the water cycle, and forests are particularly useful for regulating the way that rain is recycled back into the atmosphere to be rained down once again many miles away. By chopping down hundreds of thousands of trees every year and never replacing them, we are affecting the natural water cycle of the world, which means an increase in the pollution present in the water that now rains down on places all over the world.

- Quality of Life: The general quality of life not only for humans but also for other animals that live in, around and even many miles away from forests is gradually getting worse as a direct result of deforestation. Because soil is more often getting washed away by heavy rainfall because it is no longer anchored to tree roots, it is entering the main waterways of the world. Lakes, rivers, streams and even the sea are getting contaminated with soil that has been washed into them (which is sometimes contaminated with man-made materials such as pesticides and other chemicals). This means that creatures in the rivers and sea are in danger and also means that any animal that drinks from these water sources regularly could be in danger of getting poisoned.

- Ocean Acidification: Besides causing problems in the waterways because of an increase in soil and pesticides, for instance, deforestation has also been directly linked to ocean acidification, or the increase in the average pH of the oceans. Oceans become more acidic when more carbon dioxide is present. Because so much more carbon dioxide is now present in our air, not only because fewer trees are converting it into oxygen, but also because the processes involved in deforestation cause a lot of fossil fuels to be burned, it is gradually seeping into the oceans, raising the average pH and killing off many species of plant and animal life.

Desertification

Carbon dioxide-induced climate change and desertification remain inextricably linked because of feedbacks between land degradation and precipitation. Water resources are inextricably linked with climate. Annual average river runoff and water availability are projected to increase by 10-40% at high latitudes and in some wet tropical areas, and decrease by 10-30% over some dry regions at mid-latitudes and in the dry tropics. Soils exposed to degradation as a result of poor land management could become infertile as a result of climate change.

Climate change may exacerbate desertification through alteration of spatial and temporal patterns in temperature, rainfall, solar radiation and winds. The impacts can be described as follows:

- Soil properties and processes—including organic matter decomposition, leaching, and soil water regimes— will be influenced by temperature increase;

- At lower latitudes, especially seasonally dry and tropical regions, crop productivity is projected to decrease for even small local temperature increases (1-2°C);

- Agricultural production in many African regions is projected to be severely compromised by climate variability and change. The area suitable for agriculture, the length of growing seasons and yield potential, particularly along the margins of semi-arid and arid areas, are expected to decrease;

- In the drier areas of Latin America, climate change is expected to lead to salinisation and desertification of agricultural land;

- In Southern Europe, higher temperatures and more frequent drought are expected to reduce water availability, hydropower potential, and, in general, crop productivity.

Solutions to Desertification

The struggle against desertification can occur at several levels. Since regional variations in climate are the main causes of the loss of dryland productivity, it is important to understand the influence of global warming in specific dryland regions. According to some models of climate change, many grasslands in western North America, for example, are predicted to be at greater risk of drought due to projected increases in summer temperatures and changes to existing rainfall patterns.

At local scales, however, desertification is often the result of unsustainable land and soil management. To maintain the biological productivity of the land, soil conservation is often the priority. A number of innovative solutions have been devised that range from relatively simple changes in how people grow crops to labour-intensive landscape engineering projects. Some of the techniques that may help ameliorate the consequences of desertification in irrigated croplands, rain-fed croplands, grazing lands, and dry woodlands include:

- Salt traps, which involve the creation of so-called void layers of gravel and sand at certain depths in the soil. Salt traps prevent salts from reaching the surface of the soil and also help to inhibit water loss.

- Irrigation improvements, which can inhibit water loss from evaporation and prevent salt accumulation. This technique involves changes in the design of irrigation systems to prevent water from pooling or evaporating easily from the soil.

- Cover crops, which prevent soil erosion from wind and water. They can also reduce the local effects of drought. On larger scales, plant cover can help maintain normal rainfall patterns. Cover crops may be perennials or fast-growing annuals.

- Crop rotation, which involves the alternation of different crops on the same plot of land over different growing seasons. This technique can help maintain the productivity of the soil by replenishing critical nutrients removed during harvesting.

- Rotational grazing, which is the process of limiting the grazing pressure of livestock in a given area. Livestock are frequently moved to new grazing areas before they cause permanent damage to the plants and soil of any one area.

- Terracing, which involves the creation of multiple levels of flat ground that appear as long steps cut into hillsides. The technique slows the pace of runoff, which reduces soil erosion and retards overall water loss.

- Contour bunding (or contour bundling), which involves the placement of lines of stones along the natural rises of a landscape, and contour farming. These techniques help to capture and hold rainfall before it can become runoff. They also inhibit wind erosion by keeping the soil heavy and moist.

- Windbreaks, which involve the establishment of lines of fast-growing trees planted at right angles to the prevailing surface winds. They are primarily used to slow wind-driven soil erosion but may be used to inhibit the encroachment of sand dunes.

- Dune stabilization, which involves the conservation of the plant community living along the sides of dunes. The upper parts of plants help protect the soil from surface winds, whereas the root network below keeps the soil together.

- Charcoal conversion improvements, which include the use of steel or mud kilns or high-pressure compacting equipment to press the wood and other plant residues into briquettes. Conversion improvements retain a greater fraction of the heating potential of fuelwood.

Ozone Depletion

Principal human produced chlorine and bromine gases: Human activities cause the emission of halogen source gases that contain chlorine and bromine atoms. These emissions into the atmosphere ultimately lead to stratospheric ozone depletion. The source gases that contain only carbon, chlorine, and fluorine are called "chlorofluorocarbons," usually abbreviated as CFCs. CFCs, along with carbon tetrachloride (CCl_4) and methyl chloroform (CH_3CCl_3), historically have been the most important chlorine-containing gases that are emitted by human activities and destroy stratospheric ozone. These and other chlorine-containing gases have been used in many applications, including refrigeration, air conditioning, foam blowing, aerosol propellants, and cleaning of metals and electronic components. These activities have typically caused the emission of halogen-containing gases to the atmosphere.

Another category of halogen source gases contains bromine. The most important of these are the "halons" and methyl bromide (CH_3Br). Halons are halogenated hydrocarbon gases originally developed to extinguish fires. Halons are widely used to protect large computers, military hardware, and commercial aircraft engines. Because of these uses, halons are often directly released into the atmosphere. Halon-1211 and halon-1301 are the most abundant halons emitted by human activities. Methyl bromide, used primarily as an agricultural fumigant, is also a significant source of bromine to the atmosphere.

Human emissions of the principal chlorine- and bromine-containing gases have increased substantially since the middle of the 20th century. The result has been global ozone depletion, with the greatest losses occurring in polar regions.

Other human sources of chlorine and bromine: Other chlorine- and bromine-containing gases are released regularly in human activities. Common examples are the use of chlorine gases to disinfect swimming pools and wastewater, fossil fuel burning, and various industrial processes. These activities do not contribute significantly to stratospheric amounts of chlorine and bromine because either the global source is small or the emitted gases are short-lived (very reactive or highly soluble) and, therefore, are removed from the atmosphere before they reach the stratosphere.

Natural sources of chlorine and bromine: There are a few halogen source gases present in the stratosphere that have large natural sources. These include methyl chloride (CH_3Cl) and methyl bromide (CH_3Br), both of which are emitted by oceanic and terrestrial ecosystems. Natural sources of these two gases contribute about 17% of the chlorine currently in the stratosphere and about 30% of the bromine. Very short-lived source gases containing bromine, such as bromoform ($CHBr_3$), are also released to the atmosphere primarily from the oceans. Only a small fraction of these emissions reaches the stratosphere, because these gases are rapidly removed in the lower atmosphere. The contribution of these very short-lived gases to stratospheric bromine is estimated to be about 24%, but this has a large uncertainty. The contribution to stratospheric chlorine of short-lived chlorinated gases from natural and human sources is much smaller (< 3%) and is included in the "Other gases" category in figure. Changes in the natural sources of chlorine and bromine since the middle of the 20th century are not the cause of observed ozone depletion.

Lifetimes and emissions: After emission, halogen source gases are either naturally removed from the atmosphere or undergo chemical conversion. The time to remove or convert about 60% of a gas is often called its atmospheric "lifetime." Lifetimes vary from less than 1 year to 100 years for the principal chlorine and brominecontaining gases. Gases with the shortest lifetimes (e.g., the HCFCs, methyl bromide, methyl chloride, and the very short-lived gases) are substantially destroyed in the troposphere, and therefore only a fraction of each emitted gas contributes to ozone depletion in the stratosphere.

The amount of a halogen source gas present in the atmosphere depends on the lifetime of the gas and the amount emitted to the atmosphere. Emissions vary greatly for the principal source gases, as indicated in Table. Emissions of most gases regulated by the Montreal Protocol have decreased since 1990, and emissions from all regulated gases are expected to decrease in the coming decades.

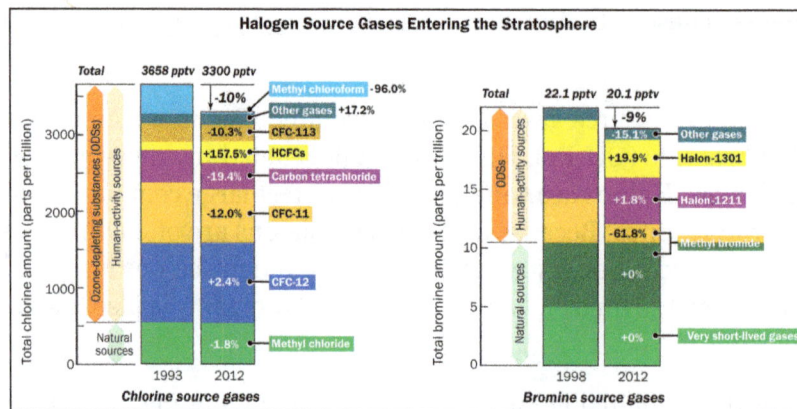

In figure above, stratospheric source gases. A variety of gases transport chlorine and bromine into the stratosphere. These gases, called halogen source gases, are emitted from natural sources and by human activities. These partitioned columns show how the principal chlorine and bromine source gases contribute to the respective total amounts of chlorine and bromine as measured in 2004. Note the large difference in the vertical scales: total chlorine in the stratosphere is 160 times more abundant than total bromine. For chlorine, human activities account for most that reaches the stratosphere. The CFCs are the most abundant of the chlorine-containing gases released in human activities. Methyl chloride is the most important natural source of chlorine. HCFCs, which are substitute gases for CFCs and also are regulated under the Montreal Protocol, are a small but growing fraction of chlorine-containing gases. The "Other gases" category includes minor CFCs and short-lived gases. For bromine that reaches the stratosphere, halons and methyl bromide are the largest sources. Both gases are released in human activities. Methyl bromide has an additional natural source. Natural sources are a larger fraction of total bromine than of total chlorine. (The unit "parts per trillion" is used here as a measure of the relative abundance of a gas in air: 1 part per trillion indicates the presence of one molecule of a gas per trillion other air molecules).

Ozone Depletion Potential: The halogen source gases in figure are also known as "ozone-depleting substances" because they are converted in the stratosphere to reactive gases containing chlorine and bromine. Some of these reactive gases participate in reactions that destroy ozone. Ozone-depleting substances are compared in their effectiveness to destroy stratospheric ozone using the "Ozone Depletion Potential" (ODP), as listed in table. A gas with a larger ODP has a greater potential to destroy ozone over its lifetime in the atmosphere. The ODP is calculated on a "per mass" basis for each gas relative to CFC-11, which has an ODP defined to be 1. Halon-1211 and halon-1301 have ODPs significantly larger than CFC-11 and most other emitted gases, because bromine is much more effective overall (about 60 times) on a per-atom basis than chlorine in chemical reactions that destroy ozone in the stratosphere. The gases with small ODP values generally have short atmospheric lifetimes or fewer chlorine and bromine atoms. The production and consumption of all principal halogen source gases by humans are regulated under the provisions of the Montreal Protocol.

Fluorine and iodine: Fluorine and iodine are also halogen atoms. Many of the source gases in figure also contain fluorine atoms in addition to chlorine or bromine. After the source gases undergo conversion in the stratosphere, the fluorine content of these gases is left in chemical forms that do not cause ozone depletion. Iodine is a component of several gases that are naturally emitted from the oceans. Although iodine can participate in ozone destruction reactions, these iodinecontaining source gases generally have very short lifetimes and, as a result, most are removed in the troposphere before they reach the stratosphere.

Table: Atmospheric lifetimes, emissions, and ozone depletion potentials of halogen source gases.

Halogen Source Gas	Atmospheric Lifetime (years)	Global Emissions in 2003 [b]	Ozone Depletion Potential (ODP) [d]
Chlorine			
CFC-12	100	101-144	1
CFC-113	85	1-15	1
CFC-11	45	60-126	1
Carbon tetrachloride			
(CCl_4)	26	58-131	0.73
HCFCs	1-26	312-403	0.02-0.12
Methyl chloroform			
(CH_3CCl_3)	5	~20	0.12
Methyl chloride	1.0	1700-13600	0.02
Bromine			
Halon-1301	65	~3	16
Halon-1211	16	7-10	7.1
Methyl bromide			
(CH_3Br)	0.7	160-200	0.51
Very short-lived gases			
(e.g., $CHBr_3$)	< 0.5	c	c

Chapter 6
Instrument to Study Climate

Diverse instruments are used for the study of climate. This chapter delves into the usage of equipment such as thermometer, barometer, hygrometer, anemometer, pyranometer, rain gauge, disdrometer, transmissometer, ceilometer, etc. to provide an extensive understanding of this subject.

Thermometer

A thermometer measures the air temperature. Most thermometers are closed glass tubes containing liquids such as alcohol or mercury. When air around the tube heats the liquid, the liquid expands and moves up the tube. A scale then shows what the actual temperature is. Temperature changes forecast weather events. Thermometers measure the changes in the temperature by using a liquid such as mercury or alcohol, normally colored red. When this liquid gets hotter it expands, and when it cools it retracts, thus the recognizable form of a thin red or silver line going up or down the thermometer. Some thermometers, called spring thermometers, measure the expansion and retraction of metal to measure the temperature. Thermometers measure the temperature in three different scales: Fahrenheit, Celsius and Kelvin, a scale normally used by scientists. The thermometer's origins trace back to Galileo who used a device he called a "thermoscope."

Meteorologists predict the Earth's weather using the world's fastest supercomputers to generate sophisticated models, along with devices within weather stations that measure variables such as temperature and pressure. One of the most important measured variables is temperature. The type of thermometer used to measure temperature varies depending on the specific weather station.

Mercury Thermometer

The mercury thermometer is a device commonly found within amateur weather stations. It consists of a glass bulb connected to a stem, in which liquid mercury is placed. As the temperature increases, thermal expansion causes the mercury volume to increase and extend along the glass tube. A scale is written on the glass tube, allowing the observer to read the temperature in Celsius or Fahrenheit. Amateurs tend to favor mercury thermometers because they are cheap and easy to handle. Their main disadvantages are a slow response time to temperature change and the need for manual reading.

Resistance Thermometer

Electrical resistance describes the process by which electrons scatter within metallic wires. Temperature drives the amount of scattering, and this property led to the development of the resistance thermometer. This device consists of metallic wire such as platinum, which is wound into a coil and mounted within a steel tube. The measured resistance is directly proportional to temperature. The coil connects to associated electronics that display the temperature on a liquid crystal display. Resistance thermometers have faster response times than their mercury counterparts and are now the standard in professional weather instrumentation since they allow the automatic logging of temperature onto a computer. The data are then transmitted to a local meteorology office headquarters for analysis.

Bimetallic Strip Thermometers

A bimetallic strip thermometer consists of two strips of different metal bonded on top of each other. Because different metals expand by different amounts, a change in temperature leads to a bending of the bimetallic strip to a significant angle. The angle of deflection is proportional to the temperature change, and hence the strips are used in combination with a dial-like scale. Bimetallic strip thermometers are used in a variety of applications, from thermostats to outdoor thermometers.

Constant Volume Thermometer

A constant volume thermometer features a bulb containing a fixed amount of gas, connected to a mercury manometer, or pressure gauge. As temperature increases, the pressure of the gas changes, and a mercury manometer measures that change. Although constant volume thermometers are not used in weather stations directly, they are among the most accurate instruments used to measure temperature and are therefore often used to calibrate more common thermometers.

Barometer

A barometer is a scientific instrument used to measure atmospheric pressure, also called barometric pressure. The atmosphere is the layers of air wrapped around the Earth. That air has a weight and presses against everything it touches as gravity pulls it to Earth. Barometers measure this pressure.

Atmospheric pressure is an indicator of weather. Changes in the atmosphere, including changes in air pressure, affect the weather. Meteorologists use barometers to predict short-term changes in the weather.

A rapid drop in atmospheric pressure means that a low-pressure system is arriving. Low pressure means that there isn't enough force, or pressure, to push clouds or storms away. Low-pressure systems are associated with cloudy, rainy, or windy weather. A rapid increase in atmospheric pressure pushes that cloudy and rainy weather out, clearing the skies and bringing in cool, dry air.

A barometer measures atmospheric pressure in units of measurement called atmospheres or bars. An atmosphere (atm) is a unit of measurement equal to the average air pressure at sea level at a temperature of 15 degrees Celsius (59 degrees Fahrenheit).

The number of atmospheres drops as altitude increases because the density of air is lower and exerts less pressure. As altitude decreases, the density of air increases, as does the number of atmospheres. Barometers have to be adjusted for changes in altitude in order to make accurate atmospheric pressure readings.

Types of Barometers

Mercury Barometer

The mercury barometer is the oldest type of barometer, invented by the Italian physicist Evangelista Torricelli in 1643. Torricelli conducted his first barometric experiments using a tube of water. Water is relatively light in weight, so a very tall tube with a large amount of water had to be used in order to compensate for the heavier weight of atmospheric pressure.

Torricelli's water barometer was more than 10 meters (35 feet) in height, which rose above the roof of his home. This odd device caused suspicion among Torricelli's neighbors, who thought he was involved in witchcraft. In order to keep his experiments more secretive, Torricelli deduced that he could create a much smaller barometer using mercury, a silvery liquid that weighs 14 times as much as water.

A mercury barometer has a glass tube that is closed at the top and open at the bottom. At the bottom of the tube is a pool of mercury. The mercury sits in a circular, shallow dish surrounding the tube. The mercury in the tube will adjust itself to match the atmospheric pressure above the dish. As the pressure increases, it forces the mercury up the tube. The tube is marked with a series of measurements that track the number of atmospheres or bars. Observers can tell what the air pressure is by looking at where the mercury stops in the barometer.

Aneroid Barometer

In 1844, the French scientist Lucien Vidi invented the aneroid barometer. An aneroid barometer has a sealed metal chamber that expands and contracts, depending on the atmospheric pressure around it. Mechanical tools measure how much the chamber expands or contracts. These measurements are aligned with atmospheres or bars.

The aneroid barometer has a circular display that indicates the present number of atmospheres, much like a clock. One hand moves clockwise or counterclockwise to point to the current number of atmospheres. The terms stormy, rain, change, fair, and dry are often written above the numbers on the dial face to make it easier for people to interpret the weather. Aneroid barometers slowly replaced mercury barometers because they were easier to use, cheaper to buy, and easier to transport since they had no liquid that could spill.

Some aneroid barometers use a mechanical tool to track the changes in atmospheric pressure over a period of time. These aneroid barometers are called barographs. Barographs are barometers connected to needles that make marks on a roll of adjacent graph paper. The barograph records the number of atmospheres on the vertical axis and units of time on the horizontal. A barograph's tracking tool will rotate, usually once every day, week, or month. The spikes in the graph show when air pressure was high or low, and how long those pressure systems lasted. A severe storm, for instance, would appear as a deep, wide dip on a barograph.

Digital Barometers

Today's digital barometers measure and display complex atmospheric data more accurately and quickly than ever before. Many digital barometers display both current barometric readings and previous 1-, 3-, 6-, and 12-hour readings in a bar chart format, much like a barograph. They also account for other atmospheric readings such as wind and humidity to make accurate weather forecasts. This data is archived and stored on the barometer and can also be downloaded onto a computer for further analysis. Digital barometers are used by meteorologists and other scientists who want up-to-date atmospheric readings when conducting experiments in the lab or out in the field.

The digital barometer is now an important tool in many of today's smartphones. This type of digital barometer uses atmospheric pressure data to make accurate elevation readings. These readings help the smartphone's GPS receiver pinpoint a location more accurately, greatly improving navigation.

Developers and researchers are also using the smartphone's crowdsourcing capabilities to make more accurate weather forecasts. Apps like PressureNet automatically collect barometric measurements from each of its users, creating a vast network of atmospheric data. This data network makes it easier and faster to map out storms as they develop, especially in areas with few weather stations.

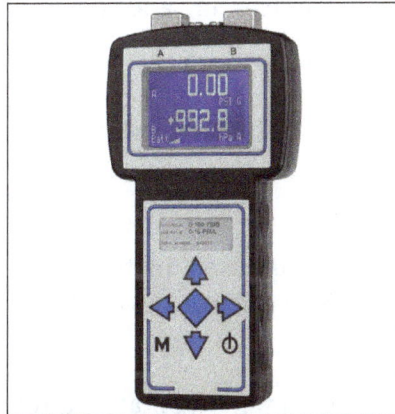

A barometer measures atmospheric pressure.

Changes in Barometric Pressure

In general, the barometer can let you know if your immediate future will see clearing or stormy skies, or you are not likely to experience a change.

- When the air is dry, cool, and pleasant, the mercury or barometer reading rises.

- When it rises, it often means clear weather.

- When the air is warm and wet, the barometer reading falls.

- When the air pressure falls, it usually indicates some type of storm or wet weather is coming.

- If the barometer remains steady, there will be no immediate change in the weather.

Predicting the Weather With the Barometer

More specifically, a barometer with readings in inches of mercury (inHg) can be interpreted in this manner:

If the reading is over 30.20 inHg (102268.9 Pa or 1022.689 mb):

- Rising or steady pressure means continued fair weather.

- Slowly falling pressure means fair weather.

- Rapidly falling pressure means cloudy and warmer conditions.

If it falls between 29.80 and 30.20 (100914.4–102268.9 Pa or 1022.689–1009.144 mb):

- Rising or steady pressure means present conditions will continue.

- Slowly falling pressure means little change in the weather.

- Rapidly falling pressure means that rain is likely, or snow if it is cold enough.

If the reading is under 29.80 (100914.4 Pa or 1009.144 mb):

- Rising or steady pressure indicates clearing and cooler weather.

- Slowly falling pressure indicates rain

- Rapidly falling pressure indicates a storm is coming.

Isobars on Weather Maps

Weather researchers (called meteorologists) use a metric unit for pressure called a millibar and they define the average pressure of a given point at sea level and 59 °F (15 °C) as one atmosphere, or 1013.25 millibars.

When a meteorologist points to a line on a weather map and refers to it as an isobar, she is referring to a line which connects points of equal atmospheric pressure. For example, a weather map will show a line connecting all points where the pressure is 996 mb (millibars) and a line below it where the pressure is 1000 mb. Points above the 1000 mb isobar have a lower pressure and points below that isobar have a higher pressure. That helps the meteorologist plot the coming changes in weather over the region.

Hygrometer

Hygrometer is an instrument used in meteorological science to measure the humidity, or amount of water vapour in the air. Several major types of hygrometers are used to measure humidity.

Mechanical hygrometers make use of the principle that organic substances (particularly finer substances such as goldbeater's skin [ox gut] and human hair) contract and expand in response to the humidity. Contraction and expansion of the hair element in a mechanical hygrometer causes the spring to move the needle on the dial.

Hygrometer.

Electrical hygrometers measure the change in electrical resistance of a thin layer of lithium chloride, or of a semiconductor device, as the humidity changes. Other hygrometers sense changes in weight, volume, or transparency of various substances that react to humidity.

Dew-point hygrometers typically consist of a polished metal mirror that is cooled at a constant pressure and constant vapour content until moisture just starts to condense on it. The temperature of the metal at which condensation begins is the dew point.

The psychrometer is a hygrometer that utilizes two thermometers—one wet-bulb and one dry-bulb—to determine humidity through evaporation. A wetted cloth wraps the wet-bulb thermometer at its enlarged end. By rapidly rotating both thermometers, or by blowing air over the bulbs, the temperature of the wet-bulb thermometer is cooler than that of the dry-bulb thermometer. The difference in temperature between the wet- and dry-bulb thermometers can be used to compute the amount of water vapour in the air.

Anemometer

An anemometer is an instrument that measures wind speed and wind pressure. Anemometers are important tools for meteorologists, who study weather patterns. They are also important to the work of physicists, who study the way air moves.

The most common type of anemometer has three or four cups attached to horizontal arms. The arms are attached to a vertical rod. As the wind blows, the cups rotate, making the rod spin. The stronger the wind blows, the faster the rod spins. The anemometer counts the number of rotations, or turns, which is used to calculate wind speed. Because wind speeds are not consistent—there are gusts and lulls—wind speed is usually averaged over a short period of time.

A similar type of anemometer counts the revolutions made by windmill-style blades. The rod of windmill anemometers rotates horizontally.

Other anemometers calculate wind speed in different ways. A hot-wire anemometer takes advantage of the fact that air cools a heated object when it flows over it. (That is why a breeze feels refreshing on a hot day.) In a hot-wire anemometer, an electrically heated, thin wire is placed in the wind. The amount of power needed to keep the wire hot is used to calculate the wind speed. The higher the wind speed, the more power is required to keep the wire at a constant temperature.

Wind speed can also be determined by measuring air pressure. (Air pressure itself is measured by an instrument called a barometer.) A tube anemometer uses air pressure to determine the wind pressure, or speed. A tube anemometer measures the air pressure inside a glass tube that is closed at one end. By comparing the air pressure inside the tube to the air pressure outside the tube, wind speed can be calculated.

Other anemometers work by measuring the speed of sound waves or by shining laser beams on tiny particles in the wind and measuring their effect.

Uses of Anemometers

Anemometers are used at almost all weather stations, from the frigid Arctic to warm equatorial regions. Wind speed helps indicate a change in weather patterns, such as an approaching storm, which is important for pilots, engineers, and climatologists.

Aerospace engineers and physicists often use laser anemometers. This type of anemometer is used in velocity experiments. Velocity is the measurement of the rate and direction of change in the position of an object. Laser anemometers calculate the wind speed around cars, airplanes, and spacecraft, for instance. Anemometers help engineers make these vehicles more aerodynamic.

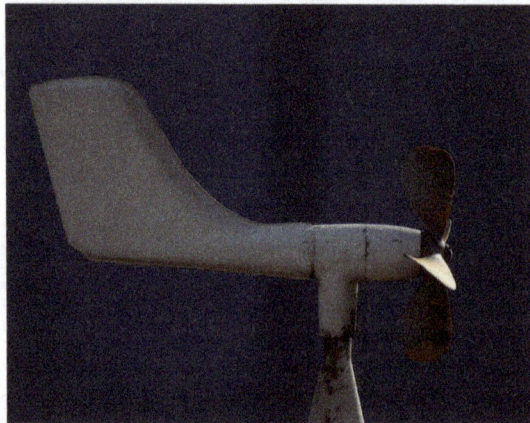

Anemometers measure wind speed and determine wind direction. Using these sets of data, meteorologists can calculate wind pressure. Wind pressure is the force exerted on a structure by the wind.

The anemometer family has many branches that use different kinds of technology to determine wind speed and pressure.

Wind Science

The unequal heating and cooling of Earth's surface cause wind. The differences in air pressure cause the air mass to shift, which results in wind that flows from high-pressure area to low-pressure areas. Around the globe, winds occur in various magnitudes including trade winds, jet streams, sea breezes and local gusts. To measure the energy borne of these winds, scientists rely on anemometers. These anemometers not only measure the current wind conditions but can also forecast potential future conditions.

Wind Speed

Several anemometers exist to measure wind speed: cup anemometers, laser doppler anemometers and sonic anemometers. Cup anemometers consist of rotating weather vanes with cups attached at the ends; the spinning rotations measure wind speed. Laser doppler anemometers use a light beam to measure the speed of particles in motion, which effectively characterizes the air speed itself. Sonic anemometers use sensors to send and receive sonic pulses across paths. The speed of the pulses can define the wind speed. Measuring wind speed is important for defining weather hazards, especially for tornado warnings and high-velocity wind exposures.

Wind Pressure

Plate anemometers help measure wind pressure. In the plate anemometer, a flat plate is compressed onto a spring, which measures the amount of force the wind exerts. Plate anemometers are mostly used in areas of high altitude. These anemometers are important during weather forecasting because they indicate times and areas of dangerous high pressure. For instance, plate anemometers are placed at bridges to raise alarms during high wind storms.

Wind Forecasting

Determining wind speed and direction provides useful information for airports, ships and everyday citizens. Crop spraying and wind farm industries rely heavily on wind patterns and use the anemometer to run their daily operations. Aircraft landing systems use anemometers to gauge their correct landing speed and protocol. Wind chill is a combination of wind speed and temperature, which results in lower temperature levels for the body.

Pyranometer

A pyranometer is a tool used to measure solar irradiance on the surface of the earth. In simple terms a pyranometer measures the amount of sunlight reaching the earth's horizontal plane. The pyranometer was invented by a Swedish meteorologist and physicist names Anders Knutsson Angstrom in 1893. His pyranometer was the first device invented that was able to measure both indirect and direct solar radiation. The main types of pyranometers are thermopile pyranometers, photodiode-based pyranometers, and photovoltaic pyranometers. These three types of pyranometers are classified into either thermopile technology or silicon semiconductor technology.

Pyranometer: A Solar Irradiance Sensor

Pyranometers measure global irradiances: the amount of solar energy per unit area per unit time incident on a surface of specific orientation emanating from a hemispherical field of view (2π sr), denoted $E_{g\downarrow}$. The global irradiance includes direct sunlight and diffuse sunlight (and in some cases specular reflections of sunlight) as illustrated in figure. The contribution from direct sunlight is given by E · cos(θ) where θ is the angle between the surface normal and the position of the sun in the sky and E is the maximum amount of direct sunlight. The global irradiance is then:

$$E_{g\downarrow} = E \cdot \cos(\theta) + E_d$$

where E_d accounts for the diffuse sunlight. In most cases the surface is horizontal such that the hemispherical field of view corresponds to the sky dome. In that case the measured quantity is the so called global horizontal irradiance (GHI) denoted $E_{g\downarrow h}$. In some cases the surface is tilted, for example in photovoltaic applications where the surface often corresponds to the plane of array (POA) of solar panels. In this case the measured quantity is the global tilted irradiance (GTI) denoted $E_{g\downarrow t}$.

A special case is the case were the surface is horizontal, but with the pyranometer facing downwards instead of towards the sky. In this case the measured quantity is the diffuse reflection from the surface of the earth, denoted $E_{r\uparrow}$.

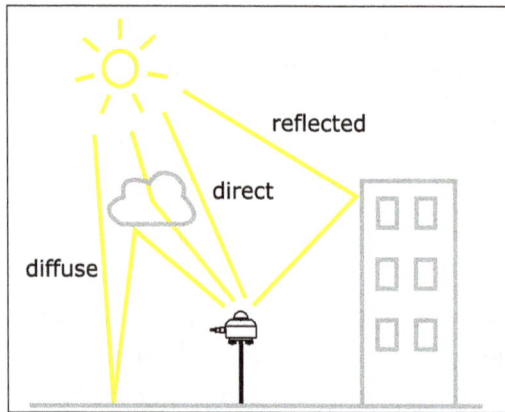

The global irradiance includes direct sunlight and diffuse sunlight and specular reflections of sunlight.

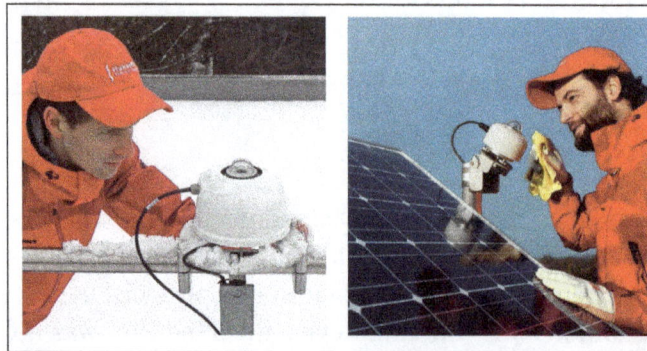

Left: a horizontally aligned pyranometer measuring the global horizontal irradiance (GHI) and right: a tilted pyranometer measuring a global tilted irradiance (GTI).

The global irradiance may vary greatly depending on the height of the sun in the sky (and thus location on the earth, time of day and time of year) and on meteorological and environmental factors such as clouds, aerosols, smog, fog, precipitation and others. Typical values for the global horizontal irradiance are in the range from 0 to 1400 W/m². In some cases it can be larger for example due to reflections from buildings or snow or in a more exotic example at the centre of a solar concentrator.

Applications of Pyranometers

The sun is earth's main source of extraterrestrial energy. This has important implications in two areas: weather and climate on the one hand and energy production by harvesting solar energy on the other hand.

Solar radiation is one of the driving forces behind the earth's weather patterns and thus an important factor in weather and climate studies. In such studies pyranometers are mostly used to measure the GHI to determine the irradiance incident on the surface of the earth. The GHI that one would measure just outside earth's atmosphere is fairly predictable, but at the surface of the earth the irradiance depends strongly on factors such as cloud coverage, aerosol concentration, fog and smog. Another interesting measurement is that of the net irradiance $E_* = E_{g\downarrow} - E_{r\uparrow}$ or the albedo $A = E_{r\uparrow} / E_{g\downarrow}$. In this case two horizontally aligned pyranometers are used: one facing towards the ground and one facing towards the sky.

In the solar energy industry pyranometers are used to monitor the performance of photovoltaic (PV) power plants. By comparing the actual power output from the PV power plant to the expected output based on a pyranometer reading the efficiency of the PV power plant can be determined. Drops in efficiency may indicate that maintenance of the PV plant is required. Pyranometers can also be used to determine the suitability of potential sites for PV power plants. In this case pyranometers are used to determine the expected output of a PV installation.

Other areas of application also exist such as building automation or agriculture.

Working of Pyraometers

Pyranometers are irradiance sensors that are based on the Seebeck- or thermoelectric effect. The main components of a pyranometer are one or two domes, a black absorber, a thermopile, the pyranometer body and in some cases additional electronics.

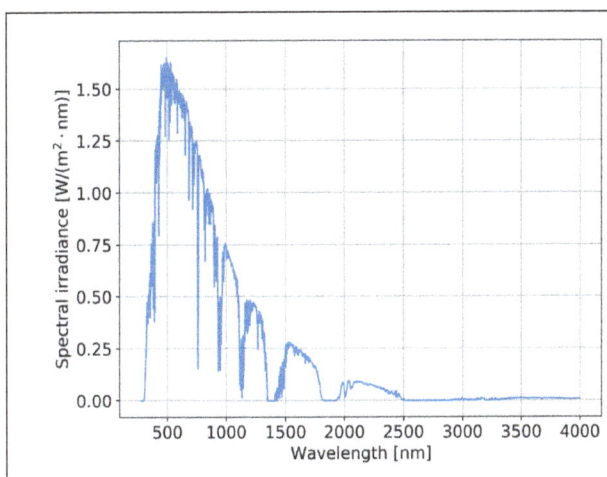

The spectral distribution of a global tilted irradiance (GTI).

The dome on a pyranometer acts as a filter that transmits solar radiation with wavelengths from roughly 300 nm to about 3 μm (this contains the near-infrared, visible, UV-A and part of the UV-B radiation), but blocks thermal radiation with wavelengths longer than 3 μm. Occasionally a second dome is used to improve the pyranometer performance. Pyranometer domes are typically made from Schott N-BK7 glass or Schott WG295 glass, but in some cases sapphire or fused silica (Spectrosil or Infrasil) domes are used. The transmission τ of solar radiation through a dome is ideally close to 100%, but is in practice closer to 92%. The dome also serves to protect the black absorber and the thermopile from the elements (rain, snow, etc.).

The filtered radiation is absorbed by the black surface on the pyranometer and converted into heat. If the transmission through the dome(s) is τ, the area of the black surface is A and the absorption coefficient of the black surface is α then the heat absorption can be calculated as follows:

$$P_{absorption} = \alpha \cdot \tau \cdot A \cdot E_{g\downarrow}$$

This creates a temperature gradient from the black surface through the thermopile to the pyranometer body which acts as a heatsink. The temperature difference is given by:

$$\Delta T = R_{thermal} \cdot P_{absorption}$$

Where $R_{thermal}$ is the thermal resistance of the thermopile sensor. This thermal resistance depends on the specific composition and geometry of the thermopile sensor. A thermopile consists of a number of thermocouples connected in series. Each thermocouple will generate a voltage proportional to the temperature difference between the black surface and the body:

$$u = \varsigma \cdot \Delta T$$

Where ς is the Seebeck coefficient. For example, the Seebeck coefficient of a copper-constantan thermocouple is 41 µV/K.

The voltage U accros the thermopile leads is simply the sum of the voltages u_i from the individual thermocouples. If the thermopile consists of N identical thermocouples the voltage across the thermopile leads is:

$$U = \sum_{i=0}^{N} u_i = N \cdot \varsigma \cdot \Delta T$$

Putting thermocouples in series allows one to detect very small temperature differences. The overall sensitivity of the pyranometer is:

$$S = \frac{U}{E} = \alpha \cdot \tau \cdot N \cdot \varsigma \cdot R_{thermal}$$

And the measured global irradiance is then:

$$E_{g\downarrow} = \frac{U}{S}$$

In practice the sensitivity is determined by calibration against a reference pyranometer rather than by calculation from the separate coefficients.

The output signal from the pyranometer can either be the output voltage from the thermopile or the pyranometer can include electronics that convert the signal from the thermopile to a more convenient output signal. Typical outputs include amplified voltage outputs, 4 -20 mA electric current outputs and digital output signals like Modbus RTU over RS-485.

The irradiance of a surface by a beam of light depends on the angle of incidence of that beam of light: the irradiance is maximum if the beam is orthogonal to that surface and zero if the beam is parallel to that surface. More generally the irradiance changes as:

$$E_{g\downarrow} = E \cdot \cos(\theta)$$

where E is the maximum irradiance (at normal incidence) and θ is the angle of incidence between the surface normal and the incident beam as illustrated in figure. Therefore the directional response of a pyranometer is this so called cosine or Lambertian response. This response is shown in figure. To get as close as possible to the Lambertian directional response pyranometers use hemispherical domes. If e.g. glass plates were used, the transmission would vary with the angle of incidence according to Fresnel's laws of optical transmission and reflection.

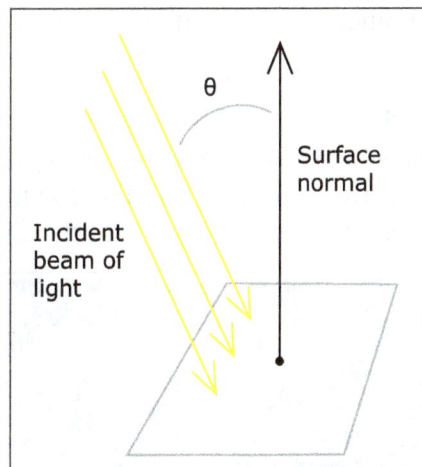

Illustration of the angle of incidence θ of a beam of light incident on a surface.

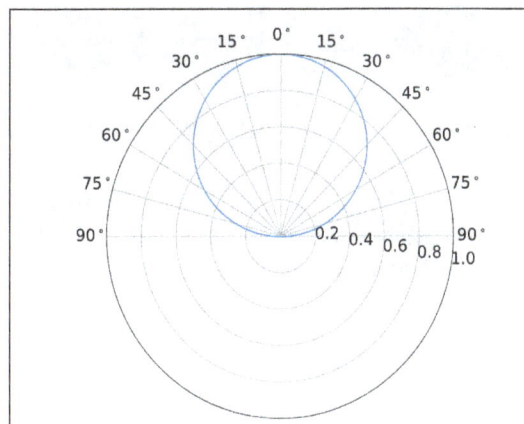

A polar plot of the Lambertian or cosine directional response (distance to origin corresponds to $E_{g\downarrow}/E = \cos(\theta)$) of a pyranometer as a function of the angle of incidence. The directional response is 1 for an angle of incidence of 0° and 0 for an angle of incidence of 90°. Pyranometers are not sensitive to light coming from the bottom.

Depending on the pyranometer specifications such as the response time, thermal offsets, non-stability, non-linearity, directional response, spectral response, temperature response and tilt response; and on the calibration method, a pyranometer may be classified either as a secondary standard, first class or second class pyranometer.

A more in-depth discussion on how pyranometers work can be found in the book by Vignola et al.

Pyranometers in the Field

Performing reliable pyranometer measurements in the field adds many practical aspects such as the pyranometer alignment, limitations on data availability due to precipitation and pyranometer maintenance and calibration.

When installing pyranometers special care should be taken of the pyranometer alignment. To measure the GHI, the pyranometer must be aligned horizontally. To this end most pyranometers come equipped with a bubble level and adjustable levelling feet or ball levelling mechanisms to align the pyranometer. In some cases the pyranometer has to be aligned in a POA to measure a GTI. Special mounting brackets exist specifically for this purpose. Some pyranometers even come with build-in tilt sensors to verify that the pyranometer remains aligned over time.

When taking measurements one should make sure the pyranometers has a clear hemispherical view of its surrounding. Snow, frost, rain, dew or dust collecting on the dome can absorb, scatter or focus radiation leading to erroneous measurements. Some pyranometers come equipped with heaters and fans to deal with snow, frost and dew and thereby increase the data availability. To avoid dust from collecting on the dome frequent cleaning of the dome is recommended.

Dew (top left), frost (top right), snow (bottom left) and rain (bottom right) can affect pyranometer readings.

Rain Gauge

The rain gauge is a meteorological instrument for measuring the amount of precipitation (especially rainfall amounts) fallen during a given time interval at a certain location. In short – the rain gauge are used to measure rainfall. It is commonly used in personal or automatic weather stations. There are different types gauges, some use direct measurement technique or others are completely automatic.

The Rain Gauge Consists of two Important Parts:

- A collector funnel;

- Mechanism to receive and measure the collected water.

The interior of the rain gauge funnel has special coating to reduce the wetting of the surface. The cone of the funnel should be deep enough, which allows the water to flow without any risk of splash. Mesh filter are used to prevent debris like leaves of bird droppings from clogging the gauge.

Place the rain collecting device at a sufficient height (usually 3 feet from the ground), and at a distance of several feet from other objects ensures that there will be no water rebound from floor or objects around.

Types of Gauges and their Working

The rain gauges were originally manual, that is to say, a meteorological technician had to come and regularly check the rainfall amounts and empty the unit – a tedious job especially with high amounts of rain. With technological advancement in the early twenty-first century, the gauges were equipped with sensors that enable electronic collection of data to be continuous and form a distance. However, manual rain gauges are reliable and accurate and are still used by amateurs or networks of volunteer observers.

There are four types of gauges:

- Graduated cylinder (called standard or direct reading gauge);

- Tipping bucket;

- Weighing gauges;

- And optical.

Graduate Cylinder

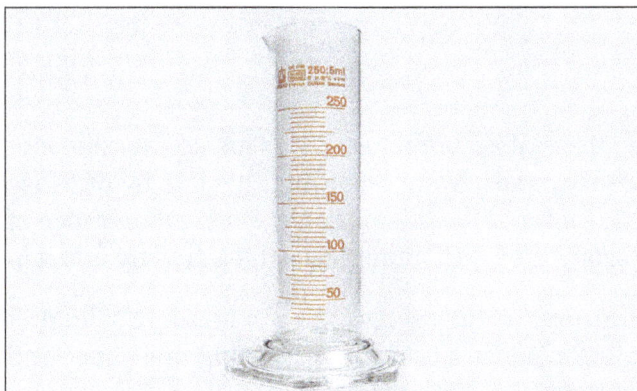

The standard rain gauge was developed at the beginning of the twentieth century and consists of a graduated cylinder (2 cm diameter) in which the collection funnel drains. Most of these gauges use scale from 0.2 mm to 25 mm. If the main unit accumulates too much water, the surplus is directed to a bigger container with diameter around 20 cm.

You record the total rainfall by measuring the total height reached on the cylinder. This is a straight forward device and quite easy to use.

Tipping Bucket

The tipping bucket gauge is made of collector funnel that directs the rain towards a two small containers, positioned on either side of a horizontal axis. The water collects inside one of the buckets that flip horizontally when it reaches the required weight, and discharges trough the force of gravity. The amount of precipitation is measured by the number of switchovers carried by the buckets, detected by a mechanical or optical system.

The advantage of this type of gauge is that it measures the rate of precipitation in addition to the total rainfall. However, when the precipitation rate is too high it may jam and report inaccurate rainfall data.

Tipping bucket rain gauge is not as precise as direct measurement, because if the rain ends before one of the buckets is full, the water inside will not be counted. Later a gust of wind can tip it and give a false accumulation when there is no rain.

The tipping bucket is the most commonly used type of rain gauge in home weather stations. Often times they use remote wireless communication to send their data.

Weighing Gauge

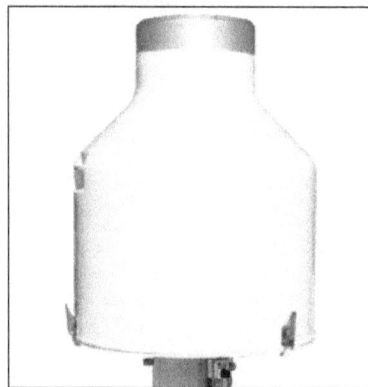

This type of rain uses the mass of accumulated water inside the collection container to calculate precipitation amount. Earlier models were recording the data by moving the tip of a stylus on to a graphic paper specially calibrated for this purpose. With the advance of technology the data is collected by a sensor and converted into numerical values directly into a logger.

This type of rain gauge measures all the rainfall and can measure solid precipitation, such as snow and hail. However, it is more expensive than direct-reading rain gauge and requires more maintenance than the tipping buckets.

Optical Gauge

The optical gauge is has a funnel on top of a photodiode or a laser diode. Rainfall is measured by detecting optical irregularities. The funnel directs the drops at the light beam, then by measuring the intensity of scintillation it can electronically determine the rate of precipitation.

Disdrometer

A disdrometer or called distrometer/rain spectrometer is a laser instrument that measures the drop size distribution falling hydrometeors.

Based on the principle of optical laser active detection, the disdrometer can continuously observe the raindrops definition size, velocity and quantity of raindrops.

It can deduce the drop size distributions, precipitation, radar reflectivity, precipitation type, etc. It can distinguish drizzle, rain, snow, hail, snow and mixed precipitation.

The instrument is sensitive and accurate, and can be used in the field for a long time without maintenance for several industries.

Disdrometer can reflect the microphysical process of precipitation. Studying raindrops spectra is helpful to understand the development and evolution of precipitation and reveal the rain generator in clouds.

Working of a Disdrometer

When the instrument works, the transmitter emits a stable red horizontal laser beam with 650 nm wavelength. The speed of light passes through the sampling area, which is detected by the laser receiver and converted into electrical signal.

The signal passes through a high-precision and stable amplification circuit, and is synchronously converted into digital signal by the A/D converter in the receiver. The signal is sent to the micro-processing controller and retrieved by the micro-processing controller.

The number of precipitation particles, particle size, falling speed and so on are obtained after sample and calculation. The statistical results are transmitted to the PC through RS-232 or RS-485 serial channel in real time.

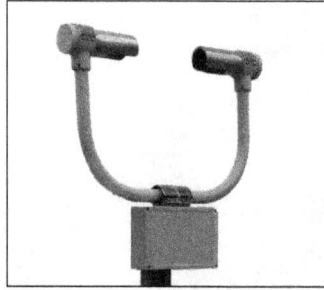

ZATA ZDM100 Disdrometer.

Measurement of raindrop sizes:

When no precipitation particles fall through the laser beam, the output signal of the receiver is the largest.

When the precipitation particles pass through the horizontal beam, they block part of the beam with their corresponding diameter, thus reducing the output signal.

We determine the diameter of precipitation particles by the size of signal reduction.

The falling velocity of precipitation particles:

The duration of the electronic signal is calculated. The duration of the electronic signal is the time that precipitation particles begin to enter the beam and leave it completely.

The following parameters can be derived from the number, size and falling velocity of precipitation particles:

- Droplet spectrum of precipitation,
- Precipitation types,
- Precipitation intensity,
- Radar reflectivity.

The splash shield installed on the laser transmitter and receiver can prevent the rain particles falling on the sensor head from bouncing back and falling into the laser beam, resulting in measurement errors.

Uses of Disdrometer

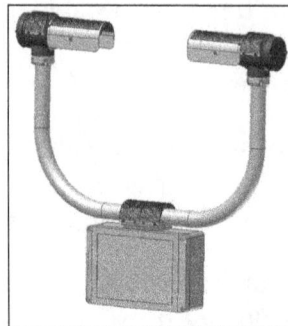

ZATA ZDM100 Disdrometer.

Application Areas:

- Traffic,

- Meteorological,

- Environmental protection,

- Hydrology.

Application Scenarios:

- Typical application in artificial rainfall work effectiveness assessment,

- Observation of raindrop spectrum and fog,

- Measurement of precipitation and fog weather phenomena,

- Zr factor calibration of rainfall radar and weather radar,

- Scientific research on precipitation process,

- Soil and water loss and soil conservation.

Transmissometer

A transmissometer is a device that is used in determining runway visual range (RVR). The transmissometer measures the extinction coefficient (the absorption of light through a medium) by using a laser shot through the atmosphere to other transmissometers at predetermined distance. The detector determines how much energy is arriving from the laser beam and from this it figures out the extinction coefficient. The transmissometer usually broadcasts a wavelength of 550 nm, which is in the middle of the visual waveband spectrum. This allows for a more then decent estimate of the RVR. They are installed on one side of the runway and normally on the ends and in the middle to provide accurate runway distance information.

A transmissometer works off the principal of transmittance. This is the fraction of like that is present after passing through a medium. For example: if you shine a flash light through a piece of dark construction paper light still shines through but it is not as bright. The transmissometer measures how much light still makes it through the atmosphere and determines how much visibility is on the runway. The worse the atmospheric conditions are the less light will shine through and the lower the runway visibility will be.

Lasers have helped make much advancement in aviation. They are used to detect clear air turbulence and wind shear. There are experiments using lasers to detect even microbursts. Since light travels so fast lasers have the ability to go for a seemingly infinite distance without converging. This allows for the exact distance of points to be measured from extremely far away. Light also has velocity so it can detect anything else moving. Lasers have the ability to do both. If air rushes through the laser a velocity can be determined. Though a transmissometer determines strength and not velocity or distance, it wouldn't be a useful tool to aviation without the help of lasers.

Ceilometer

Ceilometer is a device for measuring the height of cloud bases and overall cloud thickness. One important use of the ceilometer is to determine cloud ceilings at airports. The device works day or night by shining an intense beam of light (often produced by an infrared or ultraviolet transmitter or a laser), modulated at an audio frequency, at overhead clouds. Reflections of this light from the base of the clouds are detected by a photocell in the receiver of the ceilometer. There are two basic types of ceilometers: the scanning receiver and the rotating transmitter.

Laser Ceilometer Laser ceilometer used in the Boreal Ecosystem-Atmosphere Study (BOREAS).

The scanning-receiver ceilometer has its separate light transmitter fixed to direct its beam vertically. The receiver is stationed a known distance away. The parabolic collector of the receiver continuously scans up and down the vertical beam, searching for the point where the light intersects a cloud base. When a reflection is detected, the ceilometer measures the vertical angle to the spot; a simple trigonometric calculation then yields the height of the cloud ceiling. Many modern scanning-receiver ceilometers use a laser pulse to identify the height of a cloud's base and top and various points in between to create a vertical profile of the cloud.

The rotating-transmitter ceilometer has its separate receiver fixed to direct reflections only from directly overhead while the transmitter sweeps the sky. When the modulated beam intersects a cloud base directly over the receiver, light is reflected downward and detected.

References

- Thermometer-used-weather-stations-19389: sciencing.com, Retrieved 16 June, 2019

- Barometer: nationalgeographic.org, Retrieved 26 January, 2019

- How-to-read-a-barometer-3444043: thoughtco.com, Retrieved 15 March, 2019

- Hygrometer: britannica.com, Retrieved 19 February, 2019

- Anemometer-important-weather-forecasting- 8701078: sciencing.com, Retrieved 19 April, 2019

- Pyranometer-facts, weather-instruments – 3130: softschools.com, Retrieved 15 March, 2019

- How-to-use-rain-gauge: nwclimate.org, Retrieved 14 June, 2019

Permissions

Index

www.ingramcontent.com/pod-product-compliance
Lightning Source LLC
Chambersburg PA
CBHW061241190326
41458CB00011B/3551